CAMBRIDGE MONOGRAPHS ON MATHEMATICAL PHYSICS

General editors: P. V. Landshoff, D. W. Sciama, S. Weinberg

Scattering from black holes

SCATTERING FROM BLACK HOLES

J.A.H. FUTTERMAN, & F.A. HANDLER
Lawrence Livermore National Laboratories, Livermore, California

and

R.A. MATZNER
Department of Physics, The University of Texas at Austin

CAMBRIDGE UNIVERSITY PRESS
Cambridge New York New Rochelle
Melbourne Sydney

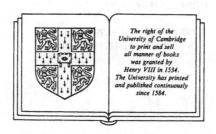

CAMBRIDGE UNIVERSITY PRESS
Cambridge, New York, Melbourne, Madrid, Cape Town, Singapore, São Paulo, Delhi

Cambridge University Press
The Edinburgh Building, Cambridge CB2 8RU, UK

Published in the United States of America by Cambridge University Press, New York

www.cambridge.org
Information on this title: www.cambridge.org/9780521112109

First published 1988
This digitally printed version 2009

A catalogue record for this publication is available from the British Library

Library of Congress Cataloguing in Publication data

Futterman, J. A. H.
Scattering from black holes.
(Cambridge monographs on mathematical physics)
Bibliography:
Includes index.
1. Black holes (Astronomy) 2. Scattering (Physics)
3. Astrophysics. I. Handler, F. A. II. Matzner,
Richard A. (Richard Alfred), 1942- . III. Title.
IV. Series.
QB843.B55F88 1987 523 86-28322

ISBN 978-0-521-32986-6 hardback
ISBN 978-0-521-11210-9 paperback

Contents

Acknowledgements

This work has benefited immeasurably from the advice of our colleagues, Cécile DeWitt-Morette, John A. Wheeler, and Bruce L. Nelson. Much of the research reported here has been supported by National Science Foundation grant #PHY81–07381 to the University of Texas. Most of the computations were done on the University of Texas Computation Center facilities, whose generous support is gratefully acknowledged. Much post-processing of the results was done on home/personal computers.

Foreword

The title says it all. Scattering, a powerful tool conceptually as well as experimentally, is applied to the simplest gravitational system, a black hole.

This study benefits gravitation, our best known and least understood phenomenon. Gravitational studies are often isolated from the mainstream of physics. This is how it should be occasionally; one needs to develop a consistent formalism *per se*; but one needs also to confront it with particle physics, cosmology, astrophysics. A knowledge of cross sections for the scattering of waves of arbitrary polarization by Schwarzschild and Kerr black holes contributes to the physical understanding of gravitation theory.

This study also benefits scattering theory. Starting with the simplest case, the scattering of massless scalar waves by a Schwarzschild black hole, the authors identify the scattering problems which have to be solved: How does one formulate scattering theory in curved spacetime? Can one define an incident plane wave in the long-range Newtonian field of a black hole? How does radiation propagate *near* a black hole? How does one handle the black hole horizon? How does one compute cross sections for polarized waves propagating in curved space-time etc...? The authors introduce several methods for solving these problems: wave mechanical scattering, partial wave decomposition, semiclassical methods, Newman–Penrose formulation of wave propagation (made powerful by Teukolsky's and Press' separation into radial polar and axial harmonics of the equation describing the evolution of wave perturbations in black hole background), and Chandrasekhar's and Detweiler's metric perturbation formalism. Numerical computations are not always of less fundamental importance than mathematical investigations; they also suggest new analytic approaches. The presence of oscillatory numerical cross sections as functions of the scattering angle, for instance, suggested an analytic calculation that identified the oscillations near the backward direction as due to glory scattering.

The scattering of massless polarized waves by Schwarzschild and Kerr black holes raises many issues which need to be confronted to be appreciated and solved consistently. This confrontation is fruitfully achieved by the authors who are skillful in all aspects of the problem, its mathematical description, its asymptotics, its approximations, its numerical computation, and its astrophysical implications.

Cécile DeWitt-Morette

1
Introduction

1.1 Motivation, context and scope

When a physicist thinks of black holes, he may think of one of two substantially different concepts. There is the astrophysical black hole, and there is the black hole of the mathematical model.

Black holes as astronomical objects are the remnants of dead stars, or perhaps one of the remnants of the inhomogeneity spectrum of the early universe. Their detection as astronomical objects has so far only been by indirect means, by observations interpreted via the astrophysicists' models. The plausible astronomical existence of black holes as X-ray sources, of black holes as the engine of quasi-stellar objects (QSO), of black holes contributing to the mass of the universe as hidden matter, makes them more interesting and more frustrating than one would expect from the mathematical description of a black hole in asymptotically flat space.

The mathematically defined black hole is the picture of simplicity. It depends only on three parameters: mass, angular momentum and charge (Schwarzschild, 1916; Reissner, 1916; Nordstrøm, 1918; Kerr, 1963, Newman *et al.*, 1965. In this work we will largely ignore charged black holes.) It is the ultimate abstraction of a physically gravitating body. One is spared the complexity of describing matter degrees of freedom, and can concentrate on the behavior of the gravitational modes.

This work treats mathematical black holes. We consider scattering of massless waves by black holes embedded in asymptotically flat spacetime. Because of the simplicity of the problem, it is to a large extent explicitly soluble; and where explicit analytic solutions are not possible, a variety of qualitative methods can be applied. Physical insight plays a large role, taking over where mathematics becomes too complicated. In the end we find the behavior of waves in black hole spacetimes; we find analogues to many classical and semiclassical scattering phenomena (e.g. Ford & Wheeler, 1959a,b), and we are able to present a fairly wide range of such phenomena under one descriptive roof. Computation finally complements analysis, and a physical system of great mathematical simplicity – in its pure form – is described.

Can the computations of gravitational scattering be applied in the realm of astrophysics? With regard to the impure, astrophysically complicated black holes, we have to admit that observation has not yet given us an absolutely convincing candidate to which we might apply these ideas.

The most plausible candidate is the X-ray source Cygnus X-1 (Hjellming, 1973; Tananbaum *et al.*, 1972). A black hole of its mass ($\sim$ 10 times solar) has a diameter of only 60 km. Cygnus X-1 is certainly a dirty astrophysical black hole. It is because of its presence in a binary system where it can interact with and accrete matter from its partner that we recognize its existence at all (Lightman & Shapiro, 1976). Because of the astrophysical environment surrounding it, we expect photons to see an object much larger than 60 km in diameter. Only gravitational radiation and neutrinos would sense the underlying geometrical structure.

Our analysis concentrates on the wave nature of the scattered field. The most distinctive wave features occur with wavelengths comparable to the size of the scatterer, which for a 10 solar mass black hole have a period of roughly a millisecond. This corresponds to the range of sensitivity of a typical 'Weber bar' (Weber, 1970). The kind of wave phenomena we find in scattering are typical diffraction-like structures which could be detectable given accurate alignment of a source and the intervening scatterer, and given successful high sensitivity detectors.

A black hole in the center of the galaxy? Perhaps, and perhaps one should take seriously the idea that focusing by a black hole may give enhanced gravitational or neutrino radiation at the Earth's surface. One may in fact ask whether the kind of QSO multi-imaging suggested by Young *et al.* (1980), due to a galaxy – or cluster – on our line of sight to the QSO could not have been instead a black hole doing the focusing. Of course it could have. However, a very massive black hole would in any case typically live in a cluster. The characteristic length scales are solar-sized, or solar system-sized, or larger-sized wavelengths, for $10^5\,M_\odot$, or $10^{10}\,M_\odot$, or $10^{14}\,M_\odot$ black holes. Quasar observations of gravitational deflection at centimeter wavelengths clearly cannot give 'fine tuned' information about black holes. Observations at wavelengths comparable to the characteristic length scale could do so. Because of the low frequencies involved, electromagnetic detection is unlikely to be possible, but gravitational wave detection should be.

For the solar-diameter wavelengths one may reasonably expect that space borne gravitational wave detectors will appear. In that case, though the matter surrounding the black hole would still complicate the pattern, the stronger specific features of black hole scattering will prevail. One of these specific features is strong backward scattering. This is impossible in a diffuse source and is analogous to, but stronger than, the feature of Rutherford scattering that led to the discovery of the nucleus; a central condensation is required for strong backward scattering of radiation. However, backward scattering may be very difficult to recognize in an astrophysical context. So one should then look for angular structure, a diffraction pattern, characteristic of a very compact scatterer. The angular cross sections computed and displayed in chapters 6 and 8 predict the diffraction pattern for gravitational radiation. The difficulty of sweeping the angle of observation when dealing with astrophysical scales is overcome by sweeping the observation in *frequency*. In a regime with strong, close spaced glory behavior (chapter 6), the scattered gravitational wave amplitude is $\tilde{\propto} J_4(\omega B \sin\theta)$ near the backward direction, where J_4 is the Bessel function of order 4, B is a fixed number, typically $B \sim 3\sqrt{3}\,M$ where M is the mass of the hole, and θ is in the backward octant. For the argument $x \gg 1$, the zeros of $J_p(x)$ are spaced $\sim\pi$ in x, or for fixed angle, $\pi/B\sin\theta$ in ω. Hence changing ω by a number of order $1/M$ will sweep the observer from a maximum to a minimum of the scattered amplitude. Given a new generation of gravitational wave detectors and a fortuitously placed source, we may finally have the option of comparing our theory to observation.

Mashoon (1974) has suggested a possible astrophysical test for scattering by black holes. We shall see that scattering from a *rotating* black hole will partially polarize the scattered radiation. Mashoon considers a spinning black hole in a 'single-line' binary. The light from the visible star will partially consist of radiation that has scattered from the black hole. If we are in the equatorial plane of the binary, and if the system is sufficiently 'clean' so that there is no other source of polarization, then we might be able to detect time varying polarization from the system. The polarization fraction depends on the total amount of radiation scattered, which in a typical binary system would be small. Only in the most extreme cases could this time varying polarization exceed $P \sim 10^{-4}$.

This is a monograph on black hole scattering. As relativists, we view the black hole as the predominant feature in our analysis. After all, wave scattering is a well known subject; see for instance the excellent text by Newton (1966). However, in combining the two disciplines, we find that one illuminates the other. To calculate the scattering of waves by black holes, you must learn the perturbation theory of black holes. You must then carry out a solution of the wave equations so obtained, wave equations that equate to zero generalizations of the flat space d'Alembertian. So a theory of the solution of such equations must be learned.

You must define the concept of 'incident plane wave' in the teeth of the long range Newtonian gravitational field of the black hole, and then discover the correct boundary conditions to impose in spacetimes with non-Euclidean geometry. Only when these points are settled can a straightforward wave scattering computation be carried out.

Once the computation has been carried out, the physicist becomes interested in the structure and detail of the scattering predictions. For him, the straightforward numerical computation may not be the most satisfactory. Understanding the details of the result is best done through an analytical calculation. Because of the complexity of the problem, it is not always possible to obtain analytic solutions, but this is almost always possible in limiting regimes, for instance, in the low energy or the high energy cases. The behavior of the scattering in such limits, where many of the specific features of the scattering are suppressed, and only the generic remain, allows full freedom to apply experience and intuition gleaned from other scattering problems in other fields of physics. In fact, the physicist may want to see these limiting solutions even before the numerical computations are carried out.

The organization of this book reflects the summary just given. The remainder of this introductory chapter gives a definition of black hole scattering, and a historical outline of its study. We then give a simple example of scattering – of massless scalar waves – in the simplest black hole spacetime. By using a decomposition into spherical modes, we give a paradigm for the numerical computation of the scattering, a preview and summary of chapters 2–5. In chapter 6 we take up the questions of approximations to scattering in limiting regimes. Here we find the s-wave dominated diffraction of the low frequency limit, and the remarkable high frequency glories. Chapter 7 presents the details of the numerical computation, and chapter 8 the results, culminating in section 8.6, 'Interpretation: glories and spirals' which connects the limiting results of chapter 6 with the numerical results of chapter 8.

In these paragraphs we have spoken indiscriminately of massless wave scattering and detection. Because it is no more difficult to include them all, we will treat scalar massless fields and electromagnetic radiation, as well as neutrino and gravitational waves; we will consider black hole scattering of massless spin 0, $\frac{1}{2}$, 1, 2 waves. Our notation will generally follow that of Misner *et al.* (1973); note in particular that we set $G = c = 1$. A bar over any object will denote complex conjugation. We use super- and subscript lower case Greek letters for spacetime indices with implied summation over repeated indices; also sub- or superscripts enclosed in (square) round brackets are to be (anti) symmetrized over.

1.2 Definition of black hole scattering

We consider the propagation of *test* radiation fields in the spacetime describing a black hole. This means that the radiation satisfies a *linear* wave equation, and that the influence of the wave on the spacetime will be ignored.

We introduce here the basic concepts of the Schwarzschild and the Kerr black holes. Although we explicitly introduce any detail we use, a familiarity with black hole physics will be an advantage in perusing this work. The metric for the simple Schwarzschild black hole is given by (1.3), and that for the Kerr black hole by (2.18).

A black hole endows spacetime with a non-Minkowskian topology. For an outside observer, one who does not choose to personally explore the interior of the black hole, there are horizons beyond which he cannot probe by any signal. Because we deal with *scattering*, our analysis will also chiefly be concerned with the region outside the outer horizons; at most, we must impose boundary conditions at the horizon radius. A black hole in fact possesses two outer horizons, a past horizon $\mathscr{H}^-$ which is the boundary from which ejecta from the black hole enters the exterior spacetime and has the Schwarzschild time coordinate $t = -\infty$, and the future horizon, $\mathscr{H}^+$, which is the boundary where all infalling radiation and observers lose contact with the exterior spacetime. Fig. 1.1 shows a *Penrose* (1968, 1969) *diagram* which describes the situation. Penrose diagrams use a variable compression of the radial coordinate to bring spatial infinity into a finite range for plotting. The horizons $\mathscr{H}^+$, $\mathscr{H}^-$ are shown as are the null infinities, $\mathscr{I}^+, \mathscr{I}^-$. These are the images of the distant past and future of the exterior region in the null direction, i.e. in the direction of massless radiation propagation. In a Penrose diagram, photons follow 45° paths, but constant radial coordinate motions appear curved.

The exterior of a black hole spacetime is the stationary causally connected part of the spacetime that includes spatial infinity $\mathscr{I}^\circ$, and future and past null infinity $\mathscr{I}^+, \mathscr{I}^-$; its inner boundary comprises the horizons $\mathscr{H}^+, \mathscr{H}^-$. (In the Schwarzschild case the horizons are the surfaces $r = 2M$, $t = \pm\infty$ where M is the mass of the hole.) (see, e.g., Hawking & Ellis, 1973, chapter 5.)

Near infinity ($\mathscr{I}^\pm$) the spacetime is asymptotically flat and wave propagation is simple. Near the horizons $\mathscr{H}^+, \mathscr{H}^-$ the propagation is also simple. At a finite radial coordinate outside the horizon is a region where the wave propagation is affected by the time independent potentials which appear in the wave equation. Thus $\mathscr{H}^\mp, \mathscr{I}^\mp$ are asymptotic regions affected by an interaction region.

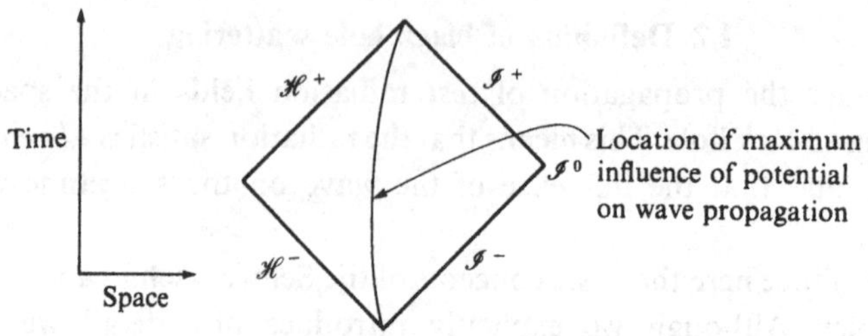

Fig. 1.1. A Penrose diagram uses a transformation of the radial coordinate which compresses spacetime near infinity to bring the region outside the horizon into a finite graph. The compression is done in such a way that radial photons follow 45° straight line paths. Constant radial coordinate lines, however, appear curved. Every point on this graph represents a 2-sphere at a fixed time and radius. Near the future and past horizon, $\mathscr{H}^+$, $\mathscr{H}^-$, and the future and past null infinities, are regions of effectively free propagation of massless radiation. The wave equation contains potentials which have maximum influence at intermediate values of the coordinate $r \sim 3M$ to $4M$ in the Schwarzschild case. In a Penrose diagram, then, there is a curved line which represents the region of the maximum influence of the potential. A straightforward scattering formulation is then possible. Figs. 1.2 and 1.3 below give diagrams of wave solutions which have 'natural' boundary conditions in the asymptotic regions of the spacetime.

We treat the scattering of massless waves from black holes in a manner completely analogous to the Schrödinger scattering of matter waves in a fixed potential field. The fact that black hole spacetimes are stationary allows us to consider eigenstates of energy, i.e. of frequency.

As in the quantum mechanical problem, we consider wave fields which solve the linearized wave equation, and which tend, in the distant past, to the form of a *plane* wave. The total solution, minus the part which is the *plane* wave is our scattered field, and is the sought-after quantity. Dividing the *energy flux* per steradian in this (*outgoing*) wave as a function of direction by the *energy flux* per unit area of the incident *plane* wave gives the differential scattering cross section.

Every concept denoted by an italicized word in the above paragraph must be carefully considered in our analysis. For instance, the concept of energy density is easily defined for scalar particles, neutrinos and photons; no strict definition of gravitational wave energy exists. (We will use an averaged, standard form due to Isaacson (1968).) The other important concept, the separation into *plane* and *outgoing* waves, is also more difficult than the short description above suggests. The gravitational interaction is a long range one. The spherically symmetric black hole is the analogue of a point charge Coulomb solution. The one characteristic feature of the long range Coulomb interaction is a logarithmic contribution to the phase of a wave field even at large distances from the scattering force center. Thus in Schrödinger scattering with Coulomb forces, one has a separation into a *distorted plane wave* and a scattered wave – both of which have the logarithmic phase contribution. We find exactly the same behavior in the gravitational case.

1.3 Historical outline

Certainly the earliest calculation of the gravitational scattering of radiation is Soldner's (1801) calculation assuming a particle theory of light – i.e. light consists of infinitesimal bullets travelling at speed c, and being deflected according to the Newtonian equation of motion. Einstein (1915) published a general relativity calculation, which for small deflections gives twice the deflection calculated in Soldner's procedure. The relativistic result has been amply verified by experiment (Jones, 1976; Reasenberg *et al.*, 1979) and is now considered one of the classical tests of relativity.

Subsequent work on *wave* as opposed to *particle* propagation has been done by Regge & Wheeler (1957), Brill & Wheeler (1957), Zerilli (1970) and Vishveshwara (1970) among others. The primary concern of these works was with the stability of the Schwarzschild black hole, and not directly with its response to externally incident wave fields. These papers, however, laid part of the basis for understanding wave behavior in black-hole backgrounds. After the discovery of the Kerr (1963) black hole, attempts were made to analyse it in the same way. (It is a non-spherical, spinning, generalization of the Schwarzschild black hole.) It leads to much more complicated wave equations, and for some time resisted attempts at a simple formulation of test field behavior.

To our knowledge, the first calculation of a *wave scattering* cross section was carried out by Hildreth (1964) who considered a scalar field interacting with a Schwarzschild black hole. Matzner (1968) considered the same problem with slightly different techniques.

Then, in 1972, Brill *et al.* separated the scalar wave equation in the Kerr geometry. Between 1972 and 1974, Press and Teukolsky, separately and together, produced a series of papers which demonstrated how to carry out perturbations in the Kerr geometry (Teukolsky, 1972, 1973; Press & Teukolsky, 1973; Teukolsky & Press, 1974). These techniques were based on work by Newman & Penrose (1962, hereafter abbreviated NP). The NP formalism concentrates on the curvature tensor and on the connection; a simple, constant form is taken for the metric. Since the curvature is *the* invariant object describing gravity, this approach naturally gives 'gauge invariant' wave quantities. Teukolsky (1973) found the remarkable result that even though the Kerr solution is only axisymmetric, a complete separation of variables is possible:

$$\Psi = e^{i\omega t} e^{im\phi} R(r) S(\theta; a\omega) \tag{1.1}$$

where the first two factors are obvious from the stationarity and the axisymmetry, but the separation of the wave into an angular part S and a radial part R was totally unexpected. This discovery meant that many technically complicated questions became very much easier to answer. In particular, the scattering problem can be solved in a separation of variables form, analogously to the separation of variables approach taken in the usual quantum mechanical problem.

For some analyses it is desirable to express electromagnetic or gravitational perturbations directly in terms of the vector potential or the perturbed metric. A general decomposition of the vector potential and metric perturbations of a rotating black hole in terms of normal modes corresponding to solutions of Teukolsky's equations was given by Chrzanowski (1975). Chrzanowski based his work on an assumption of asymptotically factorized Green's functions due to Chrzanowski & Misner (1974), and subsequently verified by Wald (1978).

Plane wave decompositions for electromagnetic and gravitational waves were given by Chrzanowski, Matzner, Sandberg & Ryan (1976, henceforth denoted by CMSR). These partial wave decompositions are valid for axial incidence only; part of the present work is thus an extension of CMSR to off axis incident waves following Futterman (1981). It is interesting to note that though CMSR and the present work make use of the vector potential and metric perturbation expansions of Chrzanowski (1975), it is not crucial to rely on that work. Use of relations between the relevant NP quantities derived by Teukolsky & Press (1974) and Starobinsky & Churilov (1973), and Chandrasekhar (1979b) allows one to work exclusively in terms of NP scalars. This is at the cost of a considerable amount of algebra and great inconvenience, however.

A great deal of analysis concerning perturbations in the Kerr metric has been carried out by Chandrasekhar (1978a,b,c, 1980), Detweiler (1976), and Chandrasekhar & Detweiler (1976, 1977). This work includes a separation of Dirac's equation in the Kerr geometry, and a transformation of Teukolsky's radial equation into a one-dimensional barrier penetration problem with a short range potential. An excellent summary is given in Chandrasekhar (1979b) along with useful information about the NP formalism, the Kerr metric, and motivation for choosing a particular NP tetrad in a given background. A qualitative description of black hole perturbation analysis is given by Detweiler (1979). The definitive discussion of the perturbations of black holes has been given by Chandrasekhar (1983).

Gravitational plane wave scattering cross sections of a Kerr black hole have been computed numerically for the case of axial incidence by Matzner & Ryan (1978), Handler (1979) and Handler & Matzner (1980). Analytical determination of low frequency limit cross sections for scalar, electromagnetic, and gravitational waves are given by Matzner & Ryan (1977) and complement numerical work, as discussed in chapter 7. Early plane wave scattering calculation was done by Matzner (1968) for massless scalar waves in the Schwarzschild geometry. (The scalar problem has been considered further, for example by Sanchez (1976, 1977), and we will quote some of her results in chapter 8.)

The original work on neutrinos in a Schwarzschild background is that of Brill & Wheeler (1957), and is extended by Rosenman (1971). Discussions of neutrinos in the Kerr geometry are given by Unruh (1973) and Martelli & Treves (1977). These last two papers discuss the absence of superradiance for neutrinos in the Kerr metric. Briefly, superradiance occurs when certain modes of integral spin (in the quantum mechanical sense) fields are amplified by carrying away some of the rotational energy of the black hole. It is the wave mechanical analogue of the Penrose (1969) process. A qualitative explanation and prediction of this effect was given by Zel'dovich (1971).

Neutrino plane waves are expanded in partial waves appropriate to the Kerr geometry in chapter 3 of the present work following the discussion in Futterman (1981). Electrons in the Kerr geometry are considered in detail by Güven (1977), following the lines of Chandrasekhar & Detweiler's (1977) analysis for neutrinos.

The references given above may be supplemented by the review of the NP formalism given by Frolov (1979). For a researcher newly entering the field of Kerr metric scattering, a basic reading list is: NP, Chandrasekhar (1983), the papers by Teukolsky & Press (1972–1974), Press & Teukolsky (1973), Matzner (1968), CMSR, Matzner & Ryan (1977, 1978), Handler & Matzner (1980) and Chandrasekhar (1979b). In addition, the work by Güven (1980) on spin-$\frac{3}{2}$ fields, and Detweiler (1980) on massive spin-0 fields near Kerr black holes will be informative. Detweiler (1982) gives a collection of selected reprints on black holes.

1.4 Partial wave analysis: the scalar wave equation in the Schwarzschild field

Many essentials of the problem are illustrated by the simplest case: the scalar (spin-0) wave in the Schwarzschild (spinless black hole) field. This special case displays the basic features directly. For more complicated (spin-$\frac{1}{2}$, -1, etc., Kerr spacetime) cases the analytical complexity can obscure the basic simplicity of the calculation. Therefore, in this section we consider massless scalar waves propagating in a Schwarzschild geometry:

$$\Box\phi \equiv (-g)^{-1/2}\partial_\mu((-g)^{1/2}g^{\mu\nu}\partial_\nu\phi) = 0. \tag{1.2}$$

Here g is the determinant of the metric tensor, whose components can be read from the line element

$$\mathrm{d}s^2 = -\mathrm{d}t^2\left(1 - \frac{2M}{r}\right) + \frac{\mathrm{d}r^2}{1 - 2M/r} + r^2\,\mathrm{d}\theta^2 + r^2\sin^2\theta\,\mathrm{d}\phi^2. \tag{1.3}$$

The *horizons* have radial coordinate $r_+ = 2M$, time coordinate $t = \pm\infty$.

Because of the stationarity and spherical symmetry of the metric, there is a natural separation:

$$\phi = \mathrm{e}^{\mathrm{i}\omega t}R_{l\omega}(r)Y_l^m(\theta, \phi),$$

where $Y_l^m(\theta, \phi)$ is a spherical harmonic, and where $R_{l\omega}(r)$ solves

$$\left\{\frac{\mathrm{d}^2}{\mathrm{d}r^{*2}} + \left[\omega^2 - \left(1 - \frac{2M}{r}\right)\left(\frac{l(l+1)}{r^2} + \frac{2M}{r^3}\right)\right]\right\}(rR_{l\omega}(r)) \tag{1.4}$$

with

$$r^* = r + 2M\ln(r/2M - 1) + \text{constant}. \tag{1.5}$$

The natural variable to use in the radial wave equation is this r^* introduced by Regge & Wheeler (1957). It clearly differs by logarithmic terms from what might be an ordinary radial variable. Notice that $r^* \to \infty$ as $r \to \infty$, but also $r^* \to -\infty$ *as* $r \to 2M$. The semi-infinite interval $r \in (2M, \infty)$ outside the horizons has been mapped into the infinite interval $r^* \in (-\infty, \infty)$. For large r, the logarithmic behavior acts exactly as does the Coulomb logarithmic phase in quantum mechanical problems, and

$$rR_{l\omega}(r) \to \mathrm{e}^{\pm i\omega r^*}$$

at $r^* \to \infty$. This logarithmic 'phase shift' at infinity is a signal that the best approximation one can make to a plane wave will be a *distorted plane wave.* The distortion arises at even very great distances from the scattering center because of the long range behavior of Newtonian gravity (or of the Coulomb interaction).

The asymptotic ($r \to \infty$) form of the distorted incident plane wave is (Matzner, 1968):

$$\phi_{\text{plane}} = \sum_l \phi_{\text{plane}}(l) \sim \sum_{l=0}^{\infty} \frac{(2l+1)}{2\mathrm{i}\omega r}[\mathrm{e}^{\mathrm{i}\omega r^*} - (-1)^l\mathrm{e}^{-\mathrm{i}\omega r^*}]P_l(\theta)\mathrm{e}^{\mathrm{i}\omega t}. \tag{1.6}$$

'down' 'up'

Fig. 1.2

This differs from what one would write in the short range case by the appearance of r^* in the exponential.

In the scattering problem one is interested in different modes which satisfy simple boundary conditions. The equation just given involves, in a straightforward sense, a superposition at infinity of ingoing and outgoing radial waves for each l. We may represent the specific solutions which have these asymptotic behaviors as (Chrzanowski, 1975) in Fig. 1.2. The 'down' solution satisfies the boundary condition that there is no radiation escaping to future null infinity $\mathscr{I}^+$. Exactly enough radiation in exactly the proper phase emerges from the past horizon $\mathscr{H}^-$ to exactly cancel any radiation that might rescatter to $\mathscr{I}^+$ from a wave that is infalling starting at the distant past at past null infinity. There is thus in this solution radiation infalling from $\mathscr{I}^-$, radiation emerging from $\mathscr{H}^-$ to meet it, and radiation infalling at $\mathscr{H}^+$ which is necessary so that 'down' is a solution to the radial wave equation. The 'up' solution is defined analogously by the boundary condition that there be no infalling radiation from $\mathscr{I}^-$, i.e. from past null infinity.

Because of the physics of the situation, which will require separate consideration of the ingoing and of the outgoing parts of the radiation at infinity, the 'up' and 'down' solutions will play an important part in what follows. For other purposes, we will later want to use an alternate decomposition in terms of modes called 'in' and 'out'.

We give an heuristic argument that the 'up', 'down' set and the 'in', 'out' set are complete sets of radial functions, at least for the scattering problem. The argument is simple. Near $\mathscr{I}^\pm$ the radial equation admits free waves, of the form $e^{\pm i\omega r^*}$. Clearly one may expand any function at $r \to \infty$ by a superposition of $e^{\pm i\omega r^*}$ waves, so the 'up', 'down' set are complete for waves that reach $\mathscr{I}^\pm$. The 'in', 'out' set are straightforward linear combinations of the 'up' 'down' set; for instance, schematically:

$$\text{'in'} = \text{'down'} - A \times \text{'up'},$$

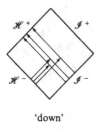

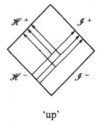

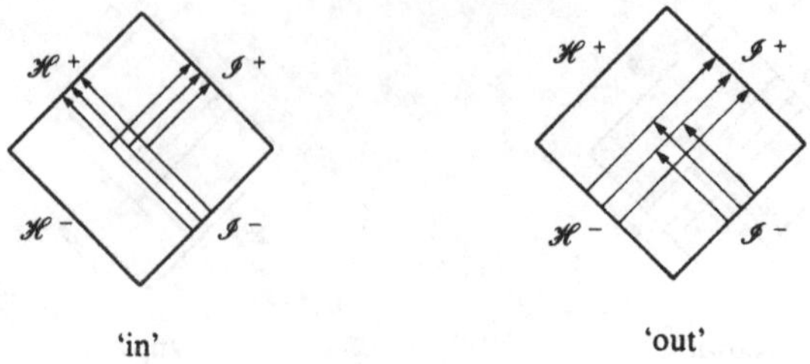

Fig. 1.3

where A is chosen to make the 'up' and 'down' waves equal on $\mathscr{H}^-$. Hence 'in' and 'out' are also complete at $\mathscr{I}^{\pm}$.

Because we know that an asymptotic plane wave will have equal amounts of ingoing and outgoing radiation at null infinity, the plane wave, ϕ_{plane}, is a superposition at infinity of 'up' and 'down' solutions. For our purposes the plane wave need not be a solution, though it can clearly be defined to be a solution by defining it equal to this superposition of 'up' and 'down' everywhere. To obtain the scattering cross section as outlined in section 1.2, one now needs an exact boundary condition. This boundary condition, appropriate to the classical (non-quantum) black holes we consider, is that the horizon is *black*; there is no field at $\mathscr{H}^-$. Call the solution satisfying this boundary condition, and normalized to be equal at $\mathscr{I}^-$ to the plane wave, $\phi(l)$; we continue to assume an angular decomposition. (Clearly therefore, $\phi(l)$ is an 'in' solution.) Then we define

$$\phi_{\text{scatt}}(l) = \phi(l) - \phi_{\text{plane}}(l). \tag{1.7}$$

Because (1.4) contains a differential operation applied to $(rR_{l\omega}(r))$, the scattered wave is asymptotically dependent on r^{-1}:

$$\phi_{\text{scatt}} \equiv \sum_l \phi_{\text{scatt}}(l) \underset{r\to\infty}{\sim} \sum_l A_l e^{i\omega r^*}/r, \tag{1.8}$$

A_l constant. Hence the flux computed by standard formulae from the stress tensor for ϕ_{scatt} is proportional to r^{-2}, and it is appropriate to compute the flux per solid angle due to ϕ_{scatt}. Dividing this quantity by the flux per unit area due to $\phi_{\text{plane}} = \sum_l \phi_{\text{plane}}(l)$ gives the differential scattering cross section.

Although this discussion has been mainly in terms of massless scalar wave fields, analogous techniques are used for any scattering problem. To continue we must develop the equations for nonzero-spin massless fields in black hole backgrounds. This will occupy us for the next few chapters.

2

Perturbations of black hole spacetimes

The scattering problem divides naturally into two major parts: the perturbation solution and the asymptology. For a given spacetime, the scattering problem is obtained by defining standard ingoing and outgoing states in the asymptotic regions of the spacetime, solving the perturbation equations, and matching (by adjusting the complex constant coefficients of the solution) the perturbation solutions to the asymptotic forms. Asymptology refers to the detailed description and normalization of the ingoing and outgoing states; it is treated in chapters 3 and 4. In this chapter we treat the derivation of the equations governing perturbations of the Kerr geometry.

The first approach taken to obtain perturbation equations was by perturbing the metric directly (in Schwarzschild) and solving for the resulting perturbed field solutions (Regge & Wheeler, 1957). This approach is the most intuitively physical approach, dealing throughout with metric quantities having direct, physical interpretations and therefore immediate connection to such quantities of interest as wave amplitudes and energy fluxes. As would be expected, the equations involved are manifestly real; in particular the scattering potentials are real. We will see that the reality of the scattering potential is intimately related to the parity of the solutions, interpretations of the wave scattering amplitudes, and numerical convenience in integrating the radial equations.

However, the above considerations are secondary in considering metric perturbations for the Kerr geometry. In Schwarzschild, direct solution for metric perturbations is a formidable task. In Kerr, the solution is very much more difficult; Chandrasekhar (1983) has given a review of this approach. Here we will concentrate on a Riemann tensor approach, based on the NP techniques. The NP formalism uses quantities which provide soluble perturbation equations, but less transparent physical interpretation. Remarkably, the perturbation equations for the Weyl tensor in the NP formalism decouple and separate (Teukolsky, 1973), leading to a partial wave analysis using suitable radial functions and spin weighted spheroidal harmonics. When this was accomplished by Teukolsky it was quite surprising since more direct approaches to the problem produced very complicated systems of equations which at that time had defied simplification. However, the NP formalism is necessarily complex in contrast to direct perturbation of the metric, which provides completely real equations. This complexity introduces numerical difficulties in solution of the radial equation, which are discussed in chapter 7.

2.1 NP formalism: perturbation of the vacuum Kerr geometry

Here we examine the equations describing a general perturbation of the Kerr background. In the following sections we will specialize to incoming plane waves. Straightforward perturbation of the metric by, say $g_{\mu\nu} = g^{\mathrm{u}}_{\mu\nu} + g^{\mathrm{p}}_{\mu\nu}$, where u and p denote unperturbed and perturbed, and expansion of the field equations, is algebraically difficult even for the Schwarzschild case.

In the NP formalism one chooses a null tetrad at every point in spacetime denoted by l, n, m, $\bar{m}$ where l and n are real and m is complex, and the bar denotes complex conjugation. The tetrad satisfies the relations

$$\begin{aligned} l_\mu l^\mu &= m_\mu m^\mu = \bar{m}_\mu \bar{m}^\mu = n_\mu n^\mu = 0, \\ l_\mu n^\mu &= -m_\mu \bar{m}^\mu = 1, \\ l_\mu m^\mu &= l_\mu \bar{m}^\mu = n_\mu m^\mu = n_\mu \bar{m}^\mu = 0. \end{aligned} \tag{2.1}$$

As can be seen from the first of (2.1), each of the vectors l, m, n is null; the other two equations of (2.1) fix the normalization.

The metric is given in terms of the tetrad by

$$g_{\mu\nu} = l_\mu n_\nu + n_\mu l_\nu - m_\mu \bar{m}_\nu - \bar{m}_\mu m_\nu. \tag{2.2}$$

Any tensor may be specified by specifying the tetrad along with all possible non-trivial projections of the tensor on the tetrad legs. Associated with a given tetrad is the connection. In the NP formalism this is specified by listing its components, the spin coefficients (Ricci rotation coefficients) conventionally denoted by κ, σ, λ, ρ, μ, τ, π, α, β, γ, ε. See Teukolsky (1973), chapter 4 below, and especially table 4.1 for the relation of these quantities to the spinor connection. These, with a given tetrad, completely specify a given spacetime (NP). (The notation for spin coefficients should be distinguished from the notation for spacetime indices.) In the case of gravitational perturbations we are interested in the scalars characterizing the Weyl tensor:

$$\begin{aligned} \Psi_0 &= C_{\alpha\beta\gamma\delta} l^\alpha m^\beta l^\gamma m^\delta, \\ \Psi_1 &= C_{\alpha\beta\gamma\delta} l^\alpha n^\beta l^\gamma m^\delta, \\ \Psi_2 &= C_{\alpha\beta\gamma\delta} l^\alpha m^\beta \bar{m}^\gamma n^\delta, \\ \Psi_3 &= C_{\alpha\beta\gamma\delta} l^\alpha n^\beta \bar{m}^\gamma n^\delta, \\ \Psi_4 &= C_{\alpha\beta\gamma\delta} n^\alpha \bar{m}^\beta n^\gamma \bar{m}^\delta. \end{aligned} \tag{2.3}$$

The NP equations are a system of first-order coupled differential equations connecting the tetrad, spin coefficients, Weyl tensor, Ricci tensor, and scalar curvature. With (2.2) a metric perturbation is equivalent to a tetrad perturbation about the background quantities

$$l = l^{\mathrm{u}} + l^{\mathrm{p}}, \quad n = n^{\mathrm{u}} + n^{\mathrm{p}}, \quad m = m^{\mathrm{u}} + m^{\mathrm{p}}. \tag{2.4}$$

The various Weyl tensor components are thereby perturbed as $\Psi = \Psi^{\mathrm{u}} + \Psi^{\mathrm{p}}$.

In the Kerr vacuum solution some simplifications are possible. Since the Kerr solution is algebraically special and of Petrov type D (Penrose, 1960; Petrov, 1955; see also Adler & Scheffield (1972)) one may choose a tetrad such that l and n lie along the repeated principal null directions of the Weyl tensor. Then from the Goldberg–Sachs (1962; see also NP) theorem we have both

$$\Psi_0{}^{\mathrm{u}} = \Psi_1{}^{\mathrm{u}} = \Psi_3{}^{\mathrm{u}} = \Psi_4{}^{\mathrm{u}} = 0 \tag{2.5}$$

and

$$\kappa^{\mathrm{u}} = \sigma^{\mathrm{u}} = \nu^{\mathrm{u}} = \lambda^{\mathrm{u}} = 0. \tag{2.6}$$

Equation (2.6) means that l and n are shear free ($\sigma = \lambda = 0$) null congruences of geodesics ($\kappa = \nu = 0$). The Weyl tetrad components Ψ_i are scalars constructed on the spacetime. Their vanishing on the background means that they are invariant under small (gauge) coordinate changes. Only Ψ_0 and Ψ_4 are invariant under small (gauge) tetrad changes. Their tetrad gauge invariance depends on the vanishing of Ψ_1 and Ψ_3 in the background. See Stewart & Walker (1973).

If the perturbation, (2.4), is carried out and use is made of (2.5) and (2.6) we find that four of the eight NP Bianchi identities and two of the 18 NP Ricci identities provide the following linear homogeneous equations to first order in the perturbations (NP; Chandrasekhar, 1978a,b,c)

$$(\bar{\delta} - 4\alpha + \pi)\Psi_0 - (\mathrm{D} - 2\varepsilon - 4\rho)\Psi_1 = 3\kappa\Psi_2, \tag{2.7}$$

$$(\Delta - 4\gamma + \mu)\Psi_0 - (\delta - 4\tau - 2\beta)\Psi_1 = 3\sigma\Psi_2, \tag{2.8}$$

$$(\mathrm{D} - \rho - \bar{\rho} - 3\varepsilon + \bar{\varepsilon})\sigma - (\delta - \tau + \bar{\pi} - \bar{\alpha} - 3\beta)\kappa = \Psi_0, \tag{2.9}$$

$$(\mathrm{D} + 4\varepsilon - \rho)\Psi_4 - (\bar{\delta} + 4\pi + 2\alpha)\Psi_3 = -3\lambda\Psi_2, \tag{2.10}$$

$$(\delta + 4\beta - \tau)\Psi_4 - (\Delta + 2\gamma + 4\mu)\Psi_3 = -3\nu\Psi_2, \tag{2.11}$$

$$(\Delta + \mu + \bar{\mu} + 3\gamma - \bar{\gamma})\lambda - (\bar{\delta} + 3\alpha + \bar{\beta} + \pi - \bar{\tau})\nu = -\Psi_4, \tag{2.12}$$

where $\Psi_0, \Psi_1, \Psi_3, \Psi_4$ and $\kappa, \sigma, \lambda, \mu$ in the above are the perturbed ($\Psi_0{}^{\mathrm{p}}$, etc) values (whose background values vanish) and all other quantities (especially Ψ_2) are unperturbed. D, δ, and Δ are intrinsic derivatives defined by

$$\mathrm{D}\phi \equiv \phi_{;\mu} l^\mu, \quad \Delta\phi \equiv \phi_{;\mu} n^\mu,$$
$$\delta\phi \equiv \phi_{;\mu} m^\mu, \quad \bar{\delta}\phi \equiv \phi_{;\mu} \bar{m}^\mu. \tag{2.13}$$

Since there are six equations for eight unknowns the solutions of the equations involve two arbitrary functions.

We will show that the equations with Ψ_0 decouple from those with Ψ_4. Consider first (2.7)–(2.9). In any type D metric (such as the Kerr solution) one may establish that (Teukolsky, 1973)

$$\bar{d}_1(\delta - 2\beta - 4\tau) = \bar{d}_2(\mathrm{D} - 2\varepsilon - 4\rho), \tag{2.14}$$

where

$$\bar{d}_1 = \mathrm{D} - 3\varepsilon + \bar{\varepsilon} - 4\rho - \bar{\rho}$$

and

$$\bar{d}_2 = \delta - 3\beta - \bar{\alpha} + \bar{\pi} - 4\tau.$$

If one operates on (2.8) with $\bar{d}_1$ and (2.7) with $\bar{d}_2$ and subtracts, the result is

$$[\bar{d}_1(\Delta - 4\gamma + \mu) - \bar{d}_2(\bar{\delta} - 4\alpha + \pi)]\Psi_0 - 3(\bar{d}_1\sigma + \bar{d}_2\kappa)\Psi_2 = 0. \tag{2.15}$$

The Bianchi identities now provide that to first order in the perturbation

$$\delta\Psi_2 = 3\tau\Psi_2 \quad \text{and} \quad \mathrm{D}\Psi_2 = 3\rho\Psi_2,$$

making the second bracket in (2.15) exactly equal to the left-hand side of (2.9). Equation (2.15) is then

$$[(\mathrm{D} - 3\varepsilon + \bar{\varepsilon} - 4\rho - \bar{\rho})(\Delta - 4\gamma + \mu)$$
$$- (\delta - 3\beta - \bar{\alpha} + \bar{\pi} - 4\tau)(\bar{\delta} - 4\alpha + \pi) - 3\Psi_2]\Psi_0 = 0. \tag{2.16}$$

Similarly, from (2.10)–(2.12) one has

$$[(\Delta + 3\gamma - \bar{\gamma} + 4\mu + \bar{\mu})(\mathrm{D} + 4\varepsilon - \rho)$$
$$- (\bar{\delta} - \bar{\tau} + \bar{\beta} + 3\alpha + 4\pi)(\delta - \tau + 4\beta) - 3\Psi_2]\Psi_4 = 0. \tag{2.17}$$

Equations (2.16) and (2.17) are the decoupled Teukolsky (1973) equations for Ψ_0 and Ψ_4.

Choosing Boyer–Lindquist (1967) coordinates, the Kerr metric is

$$\mathrm{d}s^2 = -\left(1 - \frac{2Mr}{\Sigma}\right)\mathrm{d}t^2 - \left(4Mar\frac{\sin^2\theta}{\Sigma}\right)\mathrm{d}t\,\mathrm{d}\phi$$
$$+ \left(\frac{\Sigma}{\Delta}\right)\mathrm{d}r^2 + \Sigma\,\mathrm{d}\theta^2 + \sin^2\theta\left(r^2 + a^2 + 2Ma^2r\frac{\sin^2\theta}{\Sigma}\right)\mathrm{d}\phi^2 \tag{2.18}$$

where M is the mass of the hole, aM its angular momentum,

$$\Delta = r^2 - 2Mr + a^2 \tag{2.19}$$

and

$$\Sigma = r^2 + a^2 \cos^2\theta. \tag{2.20}$$

(Do not confuse the metric function Δ with the derivative defined in (2.13).) The outer horizons of the Kerr solution have coordinates r_+ given by the larger root of $\Delta = 0$: $r_+ = M + (M^2 - a^2)^{1/2}$. In the Boyer–Lindquist coordinates the appropriate null tetrad is the Kinnersley (1969) tetrad:

$$\begin{aligned} l^\alpha &= [(r^2 + a^2)/\Delta, 1, 0, \mathrm{a}/\Delta], \\ n^\alpha &= [r^2 + a^2, -\Delta, 0, \mathrm{a}]/2(r^2 + a^2\cos^2\theta), \\ m^\alpha &= [ia\sin\theta, 0, 1, \mathrm{i}/\sin\theta]/\sqrt{2}(r + \mathrm{i}a\cos\theta), \end{aligned} \tag{2.21}$$

and the non-vanishing spin coefficients are

$$\begin{aligned} \rho &= l_{\mu;\nu} m^\mu \bar{m}^\nu = -1/(r - \mathrm{i}a\cos\theta), \\ \beta &= \tfrac{1}{2}(l_{\mu;\nu} n^\mu m^\nu - m_{\mu;\nu}\bar{m}^\mu m^\nu) = (-\bar{\rho}\cot\theta)/2\sqrt{2}, \\ \pi &= -n_{\mu;\nu}\bar{m}^\mu l^\nu = (\mathrm{i}a\rho^2 \sin\theta)/\sqrt{2}, \\ \tau &= l_{\mu;\nu} m^\mu n^\nu = (-\mathrm{i}a\rho\bar{\rho}\sin\theta)/\sqrt{2}, \\ \mu &= -n_{\mu;\nu}\bar{m}^\mu m^\nu = \rho^2\bar{\rho}\tfrac{1}{2}\Delta, \\ \gamma &= \tfrac{1}{2}(l_{\mu;\nu} n^\mu n^\nu - m_{\mu;\nu}\bar{m}^\mu n^\nu) = \mu + \rho\bar{\rho}\tfrac{1}{2}(r - M), \\ \alpha &= \tfrac{1}{2}(l_{\mu;\nu} n^\mu \bar{m}^\nu - m_{\mu;\nu}\bar{m}^\mu m^\nu) = \pi - \bar{\beta}. \end{aligned} \tag{2.22}$$

With the above, (2.16) and (2.17) may be written as a single master equation which in fact applies to the sourcefree perturbation equations for scalar, two-component neutrino, electromagnetic fields, and gravitational fields (Teukolsky, 1973):

$$\begin{aligned} &\left(\frac{(r^2 + a^2)^2}{\Delta} - a^2\sin^2\theta\right)\frac{\partial^2\Psi}{\partial t^2} + \frac{4Mar}{\Delta}\frac{\partial^2\Psi}{\partial t\partial\phi} + \left(\frac{a^2}{\Delta} - \frac{1}{\sin^2\theta}\right)\frac{\partial^2\Psi}{\partial\phi^2} \\ &\quad - \Delta^{-s}\frac{\partial}{\partial r}\left(\Delta^{s+1}\frac{\partial\Psi}{\partial r}\right) - \frac{1}{\sin\theta}\frac{\partial}{\partial\theta}\left(\sin\theta\frac{\partial\Psi}{\partial\theta}\right) - 2s\left(\frac{a(r - M)}{\Delta} + \frac{\mathrm{i}\cos\theta}{\sin^2\theta}\right)\frac{\partial\Psi}{\partial\phi} \\ &\quad - 2s\left(\frac{M(r^2 - a^2)}{\Delta} - r - \mathrm{i}a\cos\theta\right)\frac{\partial\Psi}{\partial t} + (s^2\cot^2\theta - s)\Psi = 0. \end{aligned} \tag{2.23}$$

The parameter s is called the spin weight, and will be used to label the spin of the wave under consideration. We define

$$\begin{aligned} \Psi_s &= \Psi_0 \text{ or } \rho^{-4}\Psi_4, & s &= +2 \text{ or } s = -2, \\ &= \phi_0 \text{ or } \rho^{-2}\phi_2, & s &= +1 \text{ or } s = -1, \\ &= \chi_0 \text{ or } \rho^{-1}\chi_1, & s &= \tfrac{1}{2} \text{ or } s = -\tfrac{1}{2}, \\ &= \phi, & s &= 0, \end{aligned} \tag{2.24}$$

where Ψ_0 and Ψ_4 are the Weyl components defined in (2.3), and ϕ is the scalar massless wave of chapter 1. The new variables $\phi_0, \phi_2, \chi_0, \chi_1$ represent the free components of electromagnetic radiation, and neutrino fields, respectively. The definition of the electromagnetic components parallels (2.3):

$$\begin{aligned}\phi &= F_{\alpha\beta}l^\alpha m^\beta,\\ \phi_1 &= \tfrac{1}{2}F_{\alpha\beta}(l^\alpha m^\beta + \bar{m}^\alpha n^\beta),\\ \phi_2 &= F_{\alpha\beta}\bar{m}^\alpha n^\beta,\end{aligned} \tag{2.25}$$

and the neutrino (spinor) variables are the components of the two-component Weyl spinor, expressed in the Kinnersley tetrad; cf. chapter 4.

The definitions (2.24) in (2.23) mean that the master equation (2.23) admits *separable* solutions.

$$\Psi_{(s)} = \mathrm{e}^{-\mathrm{i}\omega t}\mathrm{e}^{im\phi}{}_sS_l{}^m(\theta;a\omega)R_{l\omega}(r;\omega), \tag{2.26}$$

yielding the radial and angular equations

$$\Delta^{-s}\frac{\mathrm{d}}{\mathrm{d}r}\left(\Delta^{s+1}\frac{\mathrm{d}R}{\mathrm{d}r}\right) + \left(\frac{K^2 - 2\mathrm{i}s(r-M)K}{\Delta} + 4\mathrm{i}s\omega r - \lambda\right)R = 0 \tag{2.27}$$

and

$$\frac{1}{\sin\theta}\frac{\mathrm{d}}{\mathrm{d}\theta}\left(\sin\theta\frac{\mathrm{d}S}{\mathrm{d}\theta}\right) + \left(a^2\omega^2\cos^2\theta - \frac{m^2}{\sin^2\theta} - \frac{2ms\cos\theta}{\sin^2\theta}\right.$$
$$\left. - 2a\omega s\cos\theta - s^2\cot^2\theta + s + A\right)S = 0, \tag{2.28}$$

where

$$K = (r^2 + a^2)\omega - am, \tag{2.29a}$$

$$\lambda = A + a^2\omega^2 - 2am\omega, \tag{2.29b}$$

and s is the spin weight of the field. Equation (2.28), together with the boundary conditions of regularity at $\theta = 0$ and π, constitutes a Sturm–Liouville eigenvalue problem for the separation constant A. As such its eigenfunctions S, suitably labelled, will be orthogonal and complete (see section 2.3). We also use

$$E = A + s - s^2. \tag{2.30}$$

2.2 Relation of NP quantities to potential perturbations

The well behaved solutions $\Psi_{(s)}$ are invariant both under gauge transformations and infinitesimal tetrad rotations (thus they are measurable in principle), and uniquely and completely specify a perturbation up to changes in the parameters a and M (Wald, 1973). In principle knowledge of $\Psi_{(s)}$ suffices for formulae of most quantities of interest, in particular for the energy and angular momentum fluxes at infinity and at the horizon. Exceptions are stationary perturbations and quantum processes near the hole, for example the second quantization analysis of Unruh (1974). For physical interpretation and conceptual clarity, however, we want to work with perturbed potentials where possible. In particular, using potentials clarifies treatment of asymptotic plane waves in the mode sum approach used in chapter 3 to match the perturbation solutions of the Teukolsky equations to incident plane waves and to express the cross sections directly through the energy flux. In this section we express potential perturbations directly in terms of the NP quantities appearing in the Teukolsky equations. Since potentials only exist for integral spin perturbations, this section focuses on scalar, electromagnetic and gravitational potentials; we defer discussion of neutrino fields until chapter 4.

Since the Weyl tensor involves second derivatives of the potentials, one would expect double integration of the perturbations to be required to recover the potentials. In fact, Chrzanowski (1975) originally showed the result can be obtained more easily by assuming a factorized form for the Green function (Chrzanowski & Misner, 1974) and a later analysis by Wald (1978) confirmed Chrzanowski's result and provided a more general technique. The potentials outside the source may be written

$$P_\alpha(x) = \sum \frac{\mathrm{i}\omega}{|\omega|} g^{\mathrm{PP}'} P_\alpha^{\mathrm{up}}(x, lm\omega P)\langle P_\beta^{\mathrm{out}}(l, m, \omega, P'), S^\beta \rangle, \tag{2.31}$$

where $P_\alpha = \phi, A_\mu, h_{\mu\nu}$ is one of the perturbed potentials generated by the source terms $S_\alpha = T, J_\mu, 4T_{\mu\nu}$, and where $g^{\mathrm{PP}'}$ is the two-dimensional metric for physical polarization states, proportional to $\delta_{PP'}$ for orthogonal states. In (2.31) $\sum$ represents a sum over l, m, and P' and an integral over ω. The inner product in (2.31) is

$$\langle P_\beta, S^\beta \rangle = \int (-g)^{1/2}\,\mathrm{d}^4x \overline{P}_\beta S^\beta. \tag{2.32}$$

We use the notation up, down, in and out to denote solutions which are, respectively, vanishing on past null infinity (purely outgoing), vanishing on future null infinity (purely ingoing), vanishing on past horizon, and vanishing on the future horizon, as defined in chapter 1, figs 1.2 and 1.3. If the l, m, ω, P mode of the potential is defined as that which yields upon differentiation the l, m, ω, P mode of the field:

$${}_sD^\alpha P_\alpha^{\mathrm{up}}(l, m, \omega, P) = L(P)\Psi_{(s)}^{\mathrm{up}}(l, m, \omega) \tag{2.33}$$

Table 2.1. *Perturbed potentials and field quantities*[a]

Ingoing radiation gauge: $A_\mu l^\mu = h_{\mu\nu} l^\nu = h_\mu{}^\mu = 0$

$$\begin{aligned} A_\mu(x, lm\omega P = \pm) = {} & [-l_\mu(\bar\delta + 2\bar\beta + \bar\tau) \\ & + \bar m_\mu(\mathrm{D} + 2\bar\varepsilon + \bar\rho)]\,{}_{-1}R_{lm\omega}(r)\,{}_{+1}Z_l^m(\theta, \phi; a\omega)\mathrm{e}^{-\mathrm{i}\omega t} + P[-l_\mu(\delta + 2\beta + \tau) \\ & + m_\mu(\mathrm{D} + 2\varepsilon + \rho)]\,{}_{-1}R_{lm\omega}(r)\,{}_{-1}Z_l^m(\theta, \phi; a\omega)\mathrm{e}^{-\mathrm{i}\omega t}(-1)^{l+2m} \end{aligned}$$

$$\begin{aligned} h_{\mu\nu}(x, lm\omega P = \pm) = {} & \{-l_\mu l_\nu(\bar\delta + \alpha + 3\bar\beta - \bar\tau)(\bar\delta + 4\bar\beta + 3\bar\tau) \\ & - \bar m_\mu \bar m_\nu(\mathrm{D} - \bar\rho + 3\bar\varepsilon - \varepsilon)(\mathrm{D} + 3\bar\rho + 4\bar\varepsilon) \\ & + l_{(\mu}\bar m_{\nu)}[(\mathrm{D} + \rho - \bar\rho + \varepsilon + 3\bar\varepsilon)(\bar\delta + 4\bar\beta + 3\bar\tau) \\ & + (\bar\delta + 3\bar\beta - \alpha - \pi - \bar\tau)(\mathrm{D} + 3\bar\rho + 4\bar\varepsilon)]\}\,{}_{-2}\mathrm{R}_{lm\omega}(r)\,{}_{+2}Z_l^m(\theta, \phi; a\omega)\mathrm{e}^{-\mathrm{i}\omega t} \\ & + P\{-l_\mu l_\nu(\delta + \bar\alpha + 3\beta - \tau)(\delta + 4\beta + 3\tau) \\ & - m_\mu m_\nu(\mathrm{D} - \rho + 3\varepsilon - \bar\varepsilon)(\mathrm{D} + 3\rho + 4\varepsilon) \\ & + l_{(\mu} m_{\nu)}[(\mathrm{D} + \bar\rho - \rho + \bar\varepsilon + 3\varepsilon)(\delta + 4\beta + 3\tau) \\ & + (\delta + 3\beta - \bar\alpha - \bar\pi - \tau)(\mathrm{D} + 3\rho + 4\varepsilon)]\} \\ & \times {}_{-2}R_{lm\omega}(r)\,{}_{-2}Z_l^m(\theta, \phi; a\omega)\mathrm{e}^{-\mathrm{i}\omega t}(-1)^{l+2m}. \end{aligned}$$

Outgoing radiation gauge: $A_\mu n^\mu = h_{\mu\nu} n^\nu = h_\mu{}^\mu = 0$

$$\begin{aligned} A_\mu(x, lm\omega P = \pm) = {} & \bar\rho^{-2}[n_\mu(\delta + \bar\pi - 2\bar\alpha) - m_\mu(\Delta + \bar\mu - 2\bar\gamma)]\,{}_{+1}R_{lm\omega}(r)\,{}_{-1}Z_l^m(\theta, \phi; a\omega)\mathrm{e}^{-\mathrm{i}\omega t} \\ & + P\rho^{-2}[n_\mu(\bar\delta + \pi - 2\alpha) \\ & - \bar m_\mu(\Delta + \mu - 2\gamma)]\,{}_{+1}R_{lm\omega}(r)\,{}_{+1}Z_l^m(\theta, \phi; a\omega)\mathrm{e}^{-\mathrm{i}\omega t}(-1)^{l+2m} \end{aligned}$$

$$\begin{aligned} h_{\mu\nu}(x, lm\omega P = \pm) = {} & \bar\rho^{-4}\{-n_\mu n_\nu(\delta - 3\bar\alpha - \beta + 5\bar\pi)(\delta - 4\bar\alpha + \bar\pi) - m_\mu m_\nu(\Delta + 5\bar\mu - 3\bar\gamma + \gamma) \\ & \times (\Delta + \bar\mu - 4\bar\gamma) + n_{(\mu} m_{\nu)}[(\delta + 5\bar\pi + \beta - 3\bar\alpha + \tau)(\Delta + \bar\mu - 4\bar\gamma) \\ & + (\Delta + 5\bar\mu - \mu - 3\bar\gamma - \gamma)(\delta - 4\bar\alpha + \bar\pi)]\}\,{}_{+2}R_{lm\omega}(r)\,{}_{-2}Z_l^m(\theta, \phi; a\omega)\mathrm{e}^{-\mathrm{i}\omega t} \\ & + P\rho^{-4}\{-n_\mu n_\nu(\bar\delta - 3\alpha - \bar\beta + 5\pi)(\bar\delta - 4\alpha + \pi) \\ & - \bar m_\mu \bar m_\nu(\Delta + 5\mu - 3\gamma + \bar\gamma)(\Delta + \mu - 4\gamma) \\ & + n_{(\mu}\bar m_{\nu)}[(\bar\delta + 5\pi + \bar\beta - 3\alpha + \bar\tau)(\Delta + \mu - 4\gamma) \\ & + (\Delta + 5\mu - \bar\mu - 3\gamma - \bar\gamma)(\bar\delta - 4\alpha + \pi)]\} \\ & \times {}_{+2}R_{lm\omega}(r)\,{}_{+2}Z_l^m(\theta, \phi; a\omega)\mathrm{e}^{-\mathrm{i}\omega t}(-1)^{l+2m}. \end{aligned}$$

$$\phi_0 = (\mathrm{D} - \varepsilon + \bar\varepsilon - \bar\rho)A_m - (\delta + \bar\pi - \beta - \bar\alpha)A_l$$

$$\phi_2 = (\bar\delta + \alpha + \bar\beta - \tau)A_n - (\Delta + \bar\mu + \gamma - \bar\gamma)A_{\bar m}$$

$$\begin{aligned} -2\psi_0 = {} & (\delta + \bar\pi - 3\beta - \bar\alpha)(\delta + \bar\pi - 2\beta - 2\bar\alpha)h_{ll} + (\mathrm{D} - \bar\rho - 3\varepsilon + \bar\varepsilon)(D - \bar\rho - 2\varepsilon + 2\bar\varepsilon)h_{mm} \\ & - [(\mathrm{D} - \bar\rho - 3\varepsilon + \bar\varepsilon)(\delta + 2\bar\pi - 2\beta) + (\delta + \bar\pi - 3\beta - \bar\alpha)(\mathrm{D} - 2\bar\rho - 2\varepsilon)]h_{(lm)} \end{aligned}$$

$$\begin{aligned} -2\psi_4 = {} & (\bar\delta - \bar\tau + 3\alpha + \bar\beta)(\bar\delta - \bar\tau + 2\alpha + 2\bar\beta)h_{nn} + (\Delta + \bar\mu + 3\gamma - \bar\gamma)(\Delta + \bar\mu + 2\gamma - 2\bar\gamma)h_{\bar m\bar m} \\ & - [(\Delta + \bar\mu + 3\gamma - \bar\gamma)(\bar\delta - 2\bar\tau + 2\alpha) + (\bar\delta - \bar\tau + 3\alpha + \bar\beta)(\Delta + 2\bar\mu + 2\gamma)]h_{(n\bar m)}. \end{aligned}$$

[a]parentheses on subscripts denote symmetrization
After Chrzanowski (1975).

where ${}_sD^\alpha$ is defined in table 2.1 (see, e.g., the definition in table 2.1 of Ψ_0 in terms of h_{ll}, $h_{(l,m)}$, h_{mm}) and $L(P)$ is an amplitude factor, then acting on (2.31) with ${}_sD^\alpha$ yields

$$\Psi_{(s)} = \sum \frac{\mathrm{i}\omega}{|\omega|} \Psi_{(s)}^{\mathrm{up}}(l, m, \omega) \langle P_\beta^{\mathrm{out}}(l, m, \omega), S^\beta \rangle \tag{2.34}$$

with the definition

$$P_{\alpha}^{\text{out}}(lm\omega) = \sum_{PP'} \bar{L}(P) g^{PP'} P_{\alpha}^{\text{out}}(l, m\omega, P'). \tag{2.35}$$

The point of Chrzanowski's analysis is that (2.34) was obtained without the use of the Teukolsky equations. Hence comparison to a similar expression for $\Psi_{(s)}$ constructed from a Green function analysis of the perturbation equations allows the potentials to be read off.

Such an analysis yields (Chrzanowski, 1975)

$$\Psi_{(s)} = \sum \frac{i\omega}{|\omega|} \Psi_s^{\text{up}}(l, m, \omega) \langle {}_s R^{\text{out}}{}_s Z e^{-i\omega t}, {}_s T \rangle \tag{2.36}$$

where ${}_sT$ is the source term of the inhomogeneous Teukolsky equations (see Chrzanowski, 1975, for details). The inner product in (2.36) reveals the point of the calculation. Since the source term ${}_sT$ involves the derivatives of the sources double integration of $\Psi_{(s)}$ is equivalent to integration by parts of the inner product. The result is

$$\Psi_{(s)} = \sum \frac{i\omega}{|\omega|} \Psi_s^{\text{up}}(l, m, \omega) \langle {}_s X_{\beta}^{\text{out}}{}_s T^{\beta} \rangle, \tag{2.37}$$

where ${}_sX$ involves derivatives of ${}_sR$ and ${}_sZ$ and may be read off table 2.1 with the identity, obtained by comparing (2.34) and (2.36),

$$P_{\alpha}^{\text{out}}(l, m, \omega) = {}_s X_{\alpha}^{\text{out}}(l, m, \omega). \tag{2.38}$$

Note that these identifications are modulo gauge transformations since the source terms are arbitrary except for being divergence free. Therefore a term $\Delta_{(\alpha}\xi_{\beta)}{}^{\pm s}$ may be added to (2.31) with $\xi^{\pm s}$ two arbitrary functions, precisely the arbitrary functions mentioned concerning (2.7)–(2.12). The expressions in table 2.1 reflect a further decomposition into definite polarization states

$$P_{\alpha}(l, m, \omega, P = \pm) = P_{\alpha}(l, m, \omega) \pm \tilde{P} P_{\alpha}(l, m, \omega) \tag{2.39}$$

where $\tilde{P}$ is the parity operator $\tilde{P} = (\theta \to \pi - \theta, \phi \to \phi + \pi)$. This is the natural choice for the Kerr case since the metric is invariant under $\tilde{P}$. The resulting formulae in table 2.1 follow from the relations $\tilde{P}(\varepsilon, \rho, \mu, \gamma) = (\bar{\varepsilon}, \bar{\rho}, \bar{\mu}, \bar{\delta})$, $\tilde{P}(l_{\mu}, n_{\mu}, D, \Delta) = (l_{\mu}, n_{\mu}, D, \Delta)$ and $\tilde{P}(m, \delta, \tau, \rho, \alpha, \beta) = -(m, \mu, \delta, \rho, \alpha, \beta)$.

Chrzanowski's formulae for the potentials were later confirmed by Wald (1978) as a special case of a more general result. Wald showed that whenever a decoupled equation can be derived in a specific manner from a system of linear, partial differential equations, then a solution of the adjoint equation to the decoupled equation generates, by direct differentiation, a solution of the system of equations adjoint to the original system, hence in the self-adjoint case a solution of the original system. The result is stated succintly by Wald (1978): 'Suppose the identity $\mathscr{S}\mathscr{E} = \mathscr{O}\mathscr{T}$ holds for the linear partial operators $\mathscr{S}$, $\mathscr{E}$, $\mathscr{O}$ and $\mathscr{T}$. Suppose ψ satisfies $\mathscr{O}^\dagger\psi = 0$. Then $\mathscr{S}^\dagger\psi$ satisfies $\mathscr{E}^\dagger(\mathscr{O}^\dagger\psi) = 0$. Thus, in particular, if $\mathscr{E}$ is self-adjoint, then $\mathscr{S}^\dagger\psi$ is a solution of $\mathscr{E}(f) = 0$.' Here the operator $\mathscr{O}$ can be identified with the Teukolsky operator (left-hand side of Eqs. (2.16) and (2.17)), $\mathscr{S}$ describes the manipulations required to derive the Teukolsky equations from the field equations for the potentials, $\mathscr{T}$ is the formula for the NP components in terms of the fields, and $\mathscr{E}$ is the linear partial differential operator representing the field equation appropriate to the potential. By confirming the formulae in table 2.1, Wald confirmed the hypothesis of a factorized Green function (2.31) since Chrzanowski's argument can be reversed. Wald has also been able to generate formulae for the metric and vector potential perturbations of the Einstein–Maxwell equations in Reissner–Nordstrom spacetime (Wald, 1978).

2.3 Discussion: analytical properties of the separation functions

The functions ${}_sS_l^m(\theta;a\omega)$ of (2.26) are the spin weighted spheroidal harmonics (Teukolsky, 1972) which reduce to spherical harmonics when $a\omega = 0$ (Goldberg *et al.*, 1967). If the functions ${}_sZ_l^m(\theta,\phi;a\omega)$ are defined by

$$ {}_sZ_l^m(\theta,\phi;a\omega) = {}_sS_l^m(\theta;a\omega)e^{im\phi} \tag{2.40}$$

with the normalization

$$\int_0^{2\pi} d\phi \int_0^{\pi} d\theta \sin\theta\, {}_s\bar{Z}_l^m(\theta,\phi;a\omega)\, {}_sZ_l^m(\theta,\phi;a\omega) = 1, \tag{2.41}$$

then the decomposition (2.26) may be written

$$\Psi_{(s)} = \sum_{l,m,\omega} \psi_{(s)}(lm\omega) \sum_{l,m,\omega} {}_sR_{lm}(r)\, {}_sZ_l^m(\theta,\phi;a\omega)e^{-i\omega t}. \tag{2.42}$$

The angular equation admits among the ${}_sS_l^m$ the symmetries

$$ {}_{-s}S_l^m(\theta,a\omega) = (-1)^{l+m}\, {}_sS_l^m(\pi-\theta;a\omega), \tag{2.43}$$

where in (2.43) we take $s \geqslant 0$, and

$$ {}_sS_l^m(\theta,-a\omega) = (-1)^{l+s}\, {}_sS_l^{-m}(\pi-\theta;a\omega) \tag{2.44}$$

and in (2.44) we take $m \geqslant 0$; and among the eigenvalues

$$ {}_{-s}E_l^m(a\omega) = {}_sE_l^m(a\omega), \tag{2.45}$$

$$ {}_sE_l^m(-a\omega) = {}_sE_l^{-m}(a\omega). \tag{2.46}$$

The constants ${}_sN^m_{l;0}$, ${}_sN^m_{l;\pi}$ can be defined by the limiting behavior of the angular functions

$$ {}_sS_l^m \xrightarrow[\theta\to 0]{} \theta^{|m+s|}\,{}_sN^m_{l;0}(a\omega) \tag{2.47}$$

and

$$ {}_sS_l^m \xrightarrow[\theta\to \pi]{} (\pi-\theta)^{|m-s|}\,{}_sN^m_{l;\pi}(a\omega). \tag{2.48}$$

Equations (2.43) and (2.44) further provide

$$ {}_{-s}N^m_{l;0} = (-1)^{l+m}\,{}_sN^m_{l;\pi} \tag{2.49}$$

and

$$ {}_sN^m_{l;0}(a\omega) = (-1)^{l+s}\,{}_sN^{-m}_{l;\pi}(-a\omega). \tag{2.50}$$

It is also useful to note that (2.43) implies

$$ \tilde{P}\,{}_sZ_l{}^m(\theta,\phi;a\omega) = (-1)^{l+2m}\,{}_{-s}Z_l{}^m(\theta,\phi;a\omega), \tag{2.51a}$$

and (2.44) implies

$$ \tilde{P}\,{}_sZ_l^m(\theta,\phi;a\omega) = (-1)^{l+s+m}\,{}_sZ_l{}^{-m}(\theta,\phi;-a\omega), \tag{2.51b}$$

where $\tilde{P}$ is the parity operator $\tilde{P}$: $(\theta\to\pi-\theta,\ \phi\to\phi+\pi)$.

The symbol $P=\pm 1$ used in table 2.1 denotes the parity of the mode. It is well known (cf. Regge & Wheeler, 1957; Zerilli, 1970; CMSR) that gravitational waves of opposite parity are scattered differently by black holes. In the formalism used here, this fact appears explicitly in the normal modes of the perturbed metric. The appearance of parity in the vector potential modes is for convenience, permissible because the parity operation is a symmetry of both the Kerr metric and Maxwell's equations. In contrast, we will not decompose neutrino waves into modes of opposite parity (see chapter 4).

The factors $(-1)^{l+2m}$ appearing in table 2.1 come from (2.51). The factors $(-1)^{l+2m}$ are omitted in Chrzanowski (1975), due to the phase convention given in (2.6) of that paper, which does not reduce to the usual Condon–Shortley (1935) phase for spherical harmonics when $s=a=0$ (see Arfken, 1970). CMSR, Handler & Matzner (1980), and Futterman & Matzner (1981) use table 2.1 without the factors $(-1)^{l+2m}$, while also using ${}_sZ_l^m$ with the phase and parity behavior of (2.51) above. Their error does not affect the determination of the cross sections for fields of integer spin-weight, so no conclusions in these papers need be changed.

The behavior of the *spherical*, spin weighted angular functions

$$ {}_sY_l^m(\theta,\phi) = {}_sZ_l^m(\theta,\phi;a\omega=0) \tag{2.52}$$

has been elucidated by Goldberg *et al.* (1967). In particular, they have shown that

$$ {}_sY_l^m = {}_sS_l^m(\theta; a\omega = 0)e^{im\phi}, \tag{2.53} $$

with

$$ {}_sS_l^m(\theta; 0) = (-1)^m \left(\frac{2l+1}{4\pi}\right)^{1/2} D^l_{-sm}(\theta, \phi = 0, \psi = 0) \tag{2.54} $$

where D^l_{-sm} is a rotator function (a representation of the rotation group parameterized by the Euler angles θ, ϕ, ψ; see Appendix A, and Hammermesh, 1962, chapter 9). In general the rotator function can be written

$$ D^j_{sm}(\theta, \phi, \psi) = e^{is\psi} d^j_{sm}(\theta) e^{im\phi}. $$

Hence we also have

$$ {}_sY_l^m(\theta, \phi) = (-1)^m \left(\frac{2l+1}{4\pi}\right)^{1/2} D^l_{-sm}(\theta, \phi, 0). \tag{2.55} $$

It is unfortunate that a certain semantical loading occurs when the symbol l appears in such a formula. The standard quantum-mechanical usage for this subscript is 'j', conventionally meaning the total angular momentum which can be half integral, and must exceed both the intrinsic and orbital-z components, while in quantum mechanics, 'l' conventionally means the orbital angular momentum. From (2.53) and (2.54) we see that l occupies the place in ${}_sY_l^m$ that j would naturally occupy in quantum mechanical usage. Regardless of this psychological fact, we have that for the $s = \pm\frac{1}{2}$ case, D^j_{sm} is a 2-valued representation of the rotation group with $j \geqslant \frac{1}{2}$. Hence, when $s = \pm\frac{1}{2}$, l is a half-odd-integer in (2.55). In any case, we have

$$ l = l_{\min}, l_{\min} + 1, \cdots; \; l_{\min} = \max(|s|, |m|). \tag{2.56} $$

We note that in what follows we are led by the form of the separated equations to use the ${}_sZ_l^m(\theta, \phi; a\omega)$ as the angular basis for any spin-weight object evaluated in the Kerr geometry. We also point out that for small $a\omega$ the quantity E defined in (2.30) is

$$ E = l(l+1) + O(a\omega). \tag{2.57} $$

2.4 Asymptotic behavior, normalization and conservation relations

Because the quantities representing the wave parts of the massless fields may be expressed in terms of objects of different spin weight (e.g., in terms of ϕ_0 or ϕ_2 for electromagnetic waves, or Ψ_0, and Ψ_4 for gravitational waves) we expect relations between the functions of extreme spin weight. These relations were determined by Starobinsky & Churilov (1973), and a lucid outline of the derivation appears in Teukolsky & Press (1974). If the solutions are normalized to the positive spin solutions, the resulting relations are

$$\Psi_{(1)} = \phi_0 = {}_1R_{lm}(r,\omega)\,{}_1S_l^m(\theta; a\omega)\mathrm{e}^{-\mathrm{i}\omega t}\mathrm{e}^{\mathrm{i}m\phi}, \tag{2.58a}$$

$$\Psi_{(-1)} = \rho^{-2}\phi_2 = B\,{}_{-1}R_{lm}(r,\omega)\,{}_{-1}S_l^m(\theta; a\omega)\mathrm{e}^{-\mathrm{i}\omega t}\mathrm{e}^{\mathrm{i}m\phi}, \tag{2.58b}$$

with

$$B = [(E + a^2\omega^2 - 2a\omega m)^2 + 4am\omega - 4a^2\omega^2]^{1/2} \tag{2.59}$$

and

$$\Psi_{(2)} = \Psi_0 = {}_2R_{lm}(r,\omega)\,{}_2S_l^m(\theta; a\omega)\mathrm{e}^{-\mathrm{i}\omega t}\mathrm{e}^{\mathrm{i}m\phi} \tag{2.60a}$$

$$\Psi_{(-2)} = \rho^{-4}\Psi_4 = C\,{}_{-2}R_{lm}(r,\omega)\,{}_{-2}S_l^m(\theta; a\omega)\mathrm{e}^{-\mathrm{i}\omega t}\mathrm{e}^{\mathrm{i}m\phi} \tag{2.60b}$$

with

$$|C|^2 = (Q^2 + 4am\omega - 4a^2\omega^2)[(Q-2)^2 + 36am\omega - 36a^2\omega^2] + (2Q-1)(96a^2\omega^2 - 48am\omega) + 144\omega^2(M^2 - a^2), \tag{2.61a}$$

$$Q \equiv E + a^2\omega^2 - 2a\omega m \tag{2.61b}$$

and

$$\operatorname{Im} C = 12M\omega, \tag{2.62a}$$

$$\operatorname{Re} C = +[|C|^2 - (\operatorname{Im} C)^2]^{1/2}. \tag{2.62b}$$

The asymptotic forms of the radial equation are best seen if the radial function is transformed by

$$\mathscr{Y} = \Delta^{s/2}(r^2 + a^2)^{1/2}R \tag{2.63}$$

and if one uses the radial variable r^*, defined by

$$\frac{\mathrm{d}r^*}{\mathrm{d}r} = \frac{(r^2 + a^2)}{\Delta}. \tag{2.64}$$

The radial equation is then

$$\frac{\mathrm{d}^2\mathscr{Y}}{\mathrm{d}r^{*2}} + \left(\frac{[K^2 - 2\mathrm{i}s(r-M)K + \Delta(4\mathrm{i}r\omega s - \lambda)]}{(r^2 + a^2)^2} - G^2 - \frac{\mathrm{d}G}{\mathrm{d}r^*}\right)\mathscr{Y} = 0 \tag{2.65a}$$

with

$$G = \frac{s(r-M)}{r^2 + a^2} + \frac{r\Delta}{(r^2 + a^2)^2}. \tag{2.65b}$$

Asymptotically, (2.65a) takes the forms

$$\frac{d\mathscr{Y}}{dr^{*2}} + \left(\omega^2 + \frac{2i\omega s}{r}\right)\mathscr{Y} \sim 0 \qquad \begin{matrix} r \to +\infty \\ (r^* \to +\infty) \end{matrix} \tag{2.66}$$

and

$$\frac{d\mathscr{Y}}{dr^{*2}} + \left((\omega - m\omega_+) - \frac{is(r_+ - M)}{2Mr_+}\right)^2 \mathscr{Y} \sim 0 \qquad \begin{matrix} r \to r_+ \\ (r^* \to -\infty) \end{matrix} \tag{2.67}$$

where $\omega_+ = a/2Mr_+$ and r_+ is the event horizon $r_+ = M + (M^2 - a^2)^{1/2}$. At spatial infinity then, the asymptotic solutions are

$$\mathscr{Y} \sim r^{\pm s} e^{\mp i\omega r^*} \qquad \begin{matrix} r \to +\infty \\ (r^* \to +\infty) \end{matrix}$$

or

$$R \sim \underset{\text{(ingoing waves)}}{\frac{e^{-i\omega r^*}}{r}} \quad \text{and} \quad \underset{\text{(outgoing waves)}}{\frac{e^{i\omega r^*}}{r^{(2s+1)}}}. \tag{2.68}$$

The corresponding asymptotic behavior of the field quantities is given in table 2.2. They exhibit the peeling behavior expected of each field from the peeling theorem of Newman & Penrose (1962): for each of the fields, the ingoing and outgoing solutions differ in magnitude by the factor r^{2s}. This complicates solution of the radial equation numerically in all but the scalar and neutrino cases. In the other cases, the radial equation must be transformed for accurate solution (see section 7.2). At the horizon, the asymptotic solutions are

$$\begin{aligned} \mathscr{Y} &\sim \begin{pmatrix} Y_{\text{hole}} \\ Z_{\text{hole}} \end{pmatrix} \exp\left[\pm i\left((\omega - m\omega_+) - \frac{is(r_+ - M)}{2Mr_+}\right) r^*\right] \\ &\sim \begin{pmatrix} Y_{\text{hole}} \\ Z_{\text{hole}} \end{pmatrix} \Delta^{\pm s/2} \exp\left[\pm i(\omega - m\omega_+) r^*\right] \end{aligned} \tag{2.69a}$$

and

$$R \sim \exp\left[i(\omega - m\omega_+) r^*\right] \quad \text{or} \quad R \sim \Delta^{-s} \exp\left[-i(\omega - m\omega_+) r^*\right] \tag{2.69b}$$

for $r \to r_+$ $(r^* \to -\infty)$ where Y_{hole} is the normalizing constant for the $+s$ case, and Z_{hole} is for the $-s$ case. The normalizations of the solutions for $+s$ are related to those of $-s$ by (Teukolsky & Press, 1974):

$$BY_{\text{ingoing}} = -8\omega^2 Z_{\text{ingoing}}, \tag{2.70a}$$

$$-2\omega^2 Y_{\text{outgoing}} = BZ_{\text{outgoing}}, \tag{2.70b}$$

$$BY_{\text{hole}} = -32ikM^2 r_+{}^2(-ik + 2\varepsilon) Z_{\text{hole}}, \tag{2.70c}$$

Table 2.2. *Asymptotic behavior of the radial solutions and corresponding field quantities for $r \to \infty$ ($r^* \to \infty$)*

Field quantities	Ingoing waves	Outgoing waves	Spin
R	$\frac{1}{r}e^{-i\omega r^*}$	$e^{i\omega r^*}/r^{2s+1}$	s
ϕ	$Z_{\text{ingoing}}\frac{1}{r}e^{-i\omega r^*}$	$Z_{\text{outgoing}}\frac{1}{r}e^{i\omega r^*}$	0
χ_0	$Y_{\text{ingoing}}\frac{1}{r}e^{-i\omega r^*}$	$Y_{\text{outgoing}}\frac{1}{r^2}e^{i\omega r^*}$	$\frac{1}{2}$
χ_1	$Z_{\text{ingoing}}\frac{1}{r^2}e^{-i\omega r^*}$	$Z_{\text{outgoing}}\frac{1}{r}e^{i\omega r^*}$	$-\frac{1}{2}$
ϕ_0	$Y_{\text{ingoing}}\frac{1}{r}e^{-i\omega r^*}$	$Y_{\text{outgoing}}\frac{1}{r^3}e^{i\omega r^*}$	1
ϕ_2	$Z_{\text{ingoing}}\frac{1}{r^3}e^{-i\omega r^*}$	$Z_{\text{outgoing}}\frac{1}{r}e^{i\omega r^*}$	-1
ψ_0	$Y_{\text{ingoing}}\frac{1}{r}e^{-i\omega r^*}$	$Y_{\text{outgoing}}\frac{1}{r^5}e^{i\omega r^*}$	2
ψ_4	$Z_{\text{ingoing}}\frac{1}{r^5}e^{-i\omega r^*}$	$Z_{\text{outgoing}}\frac{1}{r}e^{i\omega r^*}$	-2

for $s = \pm 1$, where

$$\varepsilon = (M^2 - a^2)^{1/2}/4Mr_+,$$

and

$$k = \omega - m\omega^+.$$

For $s = \pm 2$, the normalization is

$$CY_{\text{ingoing}} = 64\omega^4 Z_{\text{ingoing}}, \tag{2.71a}$$

$$4\omega^4 Y_{\text{outgoing}} = \bar{C}Z_{\text{outgoing}}, \tag{2.71b}$$

$$CY_{\text{hole}} = 64(2Mr_+)^4 ik(k^2 + 4\varepsilon^2)(-ik + 4\varepsilon)Z_{\text{hole}}. \tag{2.71c}$$

We now must choose appropriate boundary conditions for the scattering problem. Since the equations are homogeneous, we may arbitrarily normalize by adjusting constant complex multipliers for each mode. The amplitudes of each l, m mode of the ingoing waves, ${}_sR_{lm}$, are matched at $r \sim \infty$ mode by mode to the amplitudes of a mode sum decomposition of an incoming plane wave. The scattering solution is defined as the one that has only ingoing waves at the horizon. In the scattering solution, then, the outgoing waves contain two pieces: the scattered and unscattered waves. The plane wave decomposition is treated in chapters 3 and 4, the scattering analysis is treated in chapter 5. Requiring the wave to be purely ingoing at the horizon sets the outgoing part of (2.69) to zero. More detail on the boundary conditions is found in chapters 3, 4, and 5.

Finally, we note that the radial equation provides for an energy conservation relation

$$\frac{dE_{\text{in}}}{dt} - \frac{dE_{\text{out}}}{dt} = \frac{dE_{\text{hole}}}{dt}, \tag{2.72}$$

where $E_{(\text{in,out,hole})}$ here stand for the energy flux into the hole from infinity due to the perturbation, the flux back out to infinity, and the energy crossing the horizon. The argument, due to Teukolsky & Press (1974), uses the fact that for an equation of the form of (2.65), the Wronskian of two solutions is independent of r. In particular, $Y(s)$ and $\bar{Y}(-s)$ are two independent solutions and therefore

$$W[Y(s), \bar{Y}(-s)]_{r=r_+} = W[Y(s), \bar{Y}(-s)]_{r=\infty}.$$

Using the asymptotic equations, (2.68) and (2.69) this yields (Teukolsky & Press, 1974) for the scalar, $s = 0$, case,

$$-4ikMr_+|Z_{\text{hole}}|^2 = -2i\omega|Z_{\text{in}}|^2 + 2i\omega|Z_{\text{out}}|^2; \tag{2.73}$$

for the electromagnetic, $s = 1$ case,

$$\frac{iB|Y_{\text{hole}}|^2}{8kMr_+} = \frac{iB|Y_{\text{in}}|^2}{4\omega} - \frac{4i\omega^3|Y_{\text{out}}|^2}{B}; \tag{2.74}$$

and for the gravitational, $s = 2$ case

$$-\frac{i\bar{C}|Y_{\text{hole}}|^2}{32k(2Mr_+)^3(k^2 + 4\varepsilon^2)} = -\frac{i\bar{C}|Y_{\text{in}}|^2}{32\omega^3} + \frac{8i\omega^5|Y_{\text{out}}|^2}{C}. \tag{2.75}$$

3

Integral spin plane waves in black hole spacetimes

This chapter gives the expansion of the incident plane waves in terms of a mode sum involving spin-weighted spheroidal harmonics. We will use Chrzanowski's (1975) normal mode expansions for vector potential and metric perturbations as summarized in table 2.1. The symbols used for the spin coefficients will occasionally be used to designate other quantities; the meaning will always be clear from the context.

3.1 Scalar waves, flat spacetime

The angular decomposition of the solution to the massless scalar wave equation

$$\Box\phi = 0 \tag{3.1}$$

for a plane wave of a particular frequency moving toward increasing z along the z-axis in flat spacetime is (Gottfried, 1966; see also chapter 1):

$$\phi_{\text{plane}} = \mathrm{e}^{-\mathrm{i}\omega t}\sum_{l=0}^{\infty}[4\pi(2l+1)]^{1/2}\mathrm{i}^{l}j_{l}(\omega r)_{0}Y_{l}^{0}(\theta,\phi) \tag{3.2}$$

where ${}_0Y_l^m$ is a spin-weight-zero spherical harmonic (${}_0Y_l^m \equiv Y_l^m$, the spherical harmonic; see Goldberg *et al.*, 1967). Here j_l is a spherical Bessel function, the solution of the flat-space radial equation:

$$\left[\frac{1}{r^2}\frac{\mathrm{d}}{\mathrm{d}r}\left(r^2\frac{\mathrm{d}}{\mathrm{d}r}\right)-\frac{l(l+1)}{r^2}+\omega^2\right]R_l(r)=0. \tag{3.3}$$

The $j_l(\omega r)$ have asymptotic behavior

$$j_l \underset{r\to\infty}{\sim} \frac{1}{2\mathrm{i}\omega r}[\exp(\mathrm{i}\omega r-\mathrm{i}\pi l/2)-\exp(-\mathrm{i}\omega r+\mathrm{i}\pi l/2)]. \tag{3.4}$$

Hence

$$\phi_{\text{plane}} \underset{r\to\infty}{\sim} \sum_{l=0}^{\infty}\frac{[4\pi(2l+1)]^{1/2}}{2\mathrm{i}\omega r}[\mathrm{e}^{\mathrm{i}\omega r}-(-1)^{l}\mathrm{e}^{-\mathrm{i}\omega r}]_{0}Y_{l}^{0}(\theta,\phi). \tag{3.5}$$

(Although we have written in a dependence the ${}_0Y_l^0$ is of course independent of ϕ).

As was mentioned in chapter 1, the long range – Newtonian – effect of the gravitational field leads to the appearance of logarithmic terms in the radial solutions. Hence we can expect these terms also in any fiducial plane wave expression. The point is that the plane wave must be an asymptotic solution; chapter 2 has shown the appearance of r^* – which contains logarithmic terms – in asymptotic solutions.

The archetype of the long-range-force scattering problem is that of a (spinless) charge in a Coulomb field. The result provides guidance for the gravitational scattering cases.

The problem of the scattering of (spinless) electrons off a point charge, as shown by Gordon (1928), has a closed-form analytic solution in parabolic cylindrical coordinates in terms of hypergeometric functions. The asymptotic form of this solution,

$$\psi \underset{r\to\infty}{\sim} \exp [\![\mathrm{i}\omega\{z - 2M \ln [\omega(r-z)]\}]\!] + \frac{f_c(\theta)}{r} \exp\{\mathrm{i}\omega[r + 2M\ln(2\omega r)]\}, \tag{3.6}$$

with

$$f_c = \frac{-M}{\sin^2(\frac{1}{2}\theta)} \exp\{2\mathrm{i}\omega M \ln[\sin^2(\tfrac{1}{2}\theta)] + \mathrm{i}\pi + 2\mathrm{i}\eta_0\} \tag{3.7}$$

and

$$\eta_0 = \arg\Gamma(1 - 2\mathrm{i}\omega M), \tag{3.8}$$

displays the existence of a logarithmic adjustment in the phase term of the second part (the 'scattered piece') as well as very distorted phase fronts in the first piece (the incident plane wave). (This solution is for the attractive scattering between a fixed-force center and a scalar electron with the parameter choice $\hbar^2/2\mu = 1$, where μ is the mass of the scattered particle, and with attractive charges of magnitude $Ze^2 = 4M\omega^2$. In this form it directly gives the Newtonian gravitational limit of scattering of a scalar massless field of frequency $|\omega|$ in a field generated by a mass M.) This identification of the two parts of (3.6) is justified by the resultant cross section using $|f_c(\theta)|^2$, which gives the experimentally measured classical Rutherford cross section (cf Messiah, 1958).

We may compare this result to that of a partial-wave decomposition. If the angular functions in the mode-sum are taken to be ${}_0Y_l^0$:

$$\psi = \sum_l u_l(r,\omega)\,{}_0Y_l^0(\theta, \phi = 0) \tag{3.9}$$

then the u_l are related to hypergeometric functions,

$$u_l = C_l r^{l+1} \mathrm{e}^{\mathrm{i}\omega r} F(l + 1 - 2\mathrm{i}M\omega, 2l + 2, -2\mathrm{i}\omega r), \tag{3.10}$$

and their asymptotic form is

$$u_l \sim r^{-1}\exp\{\pm i\omega[r + 2M\ln(2\omega r) - \tfrac{1}{2}l\pi + \eta_l]\}, \tag{3.11}$$

with

$$\eta_l = \arg\Gamma(l + 1 - 2i\omega M). \tag{3.12}$$

To solve a scattering problem from this mode-sum approach, the incident plane wave must be extracted from the full solution (3.9). Such may be done by subtracting the scattered piece, as determined by (3.6), from (3.9) provided that $f_c(\theta)$ in (3.6) be expanded in terms of spherical harmonics. As shown by Schiff (1968), the expansion of $f_c(\theta)$ is

$$f_c(\theta) = \frac{(4\pi)^{1/2}}{2i\omega}\sum_{l=0}^{\infty}(2l+1)^{1/2}{}_0Y_l^0(\theta)e^{2i\eta_l} \tag{3.13}$$

or

$$f_c(\theta) = \frac{(4\pi)^{1/2}}{2i\omega}\sum_{l=0}^{\infty}(2l+1)^{1/2}{}_0Y_l^0(\theta)(e^{2i\eta_l} - 1), \tag{3.14}$$

since $(4\pi)^{1/2}\sum_l(2l+1)^{1/2}{}_0Y_l^0(\theta) = 4\delta(1-\cos\theta)$ vanishes for $\theta \neq 0$. Hence, we find from (3.6), (3.9) and (3.14) that the asymptotic incident plane wave is

$$\frac{1}{2i\omega r}\sum_l (2l+1)[e^{i\omega_c} - (-1)^l e^{-i\omega r_c}]P_l(\theta), \tag{3.15}$$

where $r_c = r + 2M\ln 2\omega r$ and $P_l(\theta)$ is the Legendre polynomial of order l. This is identical to the asymptotic form (3.5) for a flat-space wave with the single exception that r_c replaces r in the exponents (only).

3.2 Scalar waves, spherically symmetric (Schwarzschild) spacetime

We now consider the asymptotic plane wave for scalar waves in spherically symmetric backgrounds. The Schwarzschild radial wave equation (1.4) shows that the variable $r\phi(l)$ has the asymptotic behavior $e^{\pm i\omega r^*}$. In analogy with the discussion just given, we postulate that, for the spherically symmetric case, we write

$$\phi_{\text{plane}} \underset{r\to\infty}{\sim} \frac{1}{2i\omega r}\sum (4\pi)^{1/2}(2l+1)^{1/2}[e^{i\omega r^*} - (-1)^l e^{-i\omega r^*}]_0 Y_l^0(\theta,\phi)e^{-i\omega t}. \tag{3.16}$$

The additive constant in the definition (1.5) of the Schwarzschild r^* clearly shifts the phase of the asymptotic plane wave. We will return to this point below, to exploit the freedom implied in such a choice.

3.3 Scalar waves, Kerr background (on-axis)

There is an obvious, essential difference between the angular decomposition of an asymptotic plane wave in flat space or in a spherically symmetric spacetime, and in the Kerr background. The Kerr background possesses a physically defined axis. Hence 'on-axis' scattering in spherically symmetric spacetimes is completely general, while a distinction must be made between the simpler on-axis case, and the off-axis case, for Kerr spacetimes.

One might think that the Kerr on axis case can be obtained simply by inserting the relevant spheroidal angular function, and the relevant definition of r^* into expression (3.16). However, some subtlety is required. We use Boyer–Lindquist (1967) coordinates, and

$$\phi = \mathrm{e}^{i\omega r\cos\theta}\mathrm{e}^{-i\omega t} = \sum_l {}_0C_{l0}\,{}_0Z_l^0(\theta; a\omega)$$

where we explicitly use the axisymmetry. (Notice that ${}_0C_{l0}$ is a function of t and r.)

Now

$$_0C_{l0} = \mathrm{e}^{-i\omega t}\int \mathrm{d}\Omega\,\mathrm{e}^{i\omega r\cos\theta}\,{}_0\bar{Z}_l^0(\theta; a\omega).$$

Evaluation of this integral in the large ωr limit, by the technique of stationary phase gives (Appendix A) – after substituting r^* for r in the exponents:

$$_0C_{l0} = \frac{2\pi}{i\omega r}\mathrm{e}^{-i\omega t}\,{}_0N^0_{l;0}[\mathrm{e}^{i\omega r^*} - (-1)^l\mathrm{e}^{-i\omega r^*}], \tag{3.17}$$

where we used also (2.49), and the Kerr definition of r^*:

$$\frac{\mathrm{d}r^*}{\mathrm{d}r} = \frac{r^2 + a^2}{\Delta}. \tag{3.18}$$

Now

$$_0Z_l^m(\theta, \phi; a\omega) \equiv {}_0S_l^m(\theta; a\omega)\mathrm{e}^{im\phi} \tag{3.19}$$

is a spin-weight-zero spheroidal harmonic; it is in fact identically a spheroidal harmonic in the definition of Flammer (1957), and it reduces to a spherical harmonic in the limit $a\omega \to 0$. Further, for the spherical case,

$$_0N^0_{l;0}(a\omega = 0) = \left(\frac{2l+1}{4\pi}\right)^{1/2}$$

and we regain the spherical result (3.16). For general $a\omega$, a closed form for ${}_0N^0_{l;0}$ is not possible.

3.4 Scalar waves, Kerr background, arbitrary incidence

Our approach here will be to compute the angular decomposition in terms of spheroidal functions for an off-axis plane wave in flat space and then introduce r^* for r as in the procedures described above.

Equation (3.1) admits the solution

$$\phi = b \exp[-\mathrm{i}\omega(t - r\sin\gamma\sin\theta\sin\phi - r\cos\gamma\cos\theta)], \qquad (3.20)$$

which represents a wave of frequency ω and amplitude b travelling along a ray inclined by an angle γ with respect to the z-axis. As in section 3.3, ϕ may be expanded in oblate spheroidal harmonics of spin-weight zero:

$$\phi = \sum_{lm} {}_0C_{lm}\,{}_0Z_l^m(\theta, \phi; a\omega). \qquad (3.21)$$

Where now we must include nonzero m, because of the explicit ϕ-dependence.

Again, as in section 3.3:

$${}_0C_{lm} = b\int \mathrm{d}\Omega\, \mathrm{e}^{-\mathrm{i}\chi}\,{}_0\bar{Z}_l^m(\theta, \phi; a\omega), \qquad (3.22)$$

where

$$\chi = \omega(t - r\sin\gamma\sin\theta\sin\phi - r\cos\gamma\cos\theta). \qquad (3.23)$$

The large-r limit of the integral is evaluated in Appendix A, (cf (A11), (A12)) with the result after substituting r^* for r in the exponents

$${}_0C_{lm} \underset{r\to\infty}{\sim} \frac{2\pi b}{\mathrm{i}\omega r}[(-\mathrm{i})^m \mathrm{e}^{-\mathrm{i}\omega(t-r^*)}\,{}_0S_l^m(\gamma; a\omega) - (\mathrm{i})^m \mathrm{e}^{-\mathrm{i}\omega(t+r^*)}\,{}_0S_l^m(\pi - \gamma; a\omega)]. \qquad (3.24)$$

Equation (3.24) is valid only asymptotically. Matching it to 'normal mode' decompositions valid at all radii

$$\phi^{\mathrm{up}} = \int \mathrm{d}\tilde{\omega} K^{\mathrm{up}}_{lm\tilde{\omega}}\,{}_0R^{\mathrm{up}}_{lm\tilde{\omega}}(\mathrm{r})\,{}_0Z_l^m(\theta, \phi; a\tilde{\omega})\mathrm{e}^{-\mathrm{i}\tilde{\omega}t},$$

$$\phi^{\mathrm{dn}} = \int \mathrm{d}\tilde{\omega} K^{\mathrm{dn}}_{lm\tilde{\omega}}\,{}_0R^{\mathrm{dn}}_{lm\tilde{\omega}}(\mathrm{r})\,{}_0Z_l^m(\theta, \phi; a\tilde{\omega})\mathrm{e}^{-\mathrm{i}\tilde{\omega}t}, \qquad (3.25)$$

(where up and dn refer to the boundary behavior discussed in chapter 2) yields

$$K^{\mathrm{up}}_{lm\tilde{\omega}} = \frac{2\pi b}{\mathrm{i}\omega}(-\mathrm{i})^m\,{}_0S_l^m(\gamma; a\tilde{\omega})\delta(\omega - \tilde{\omega}),$$

$$K^{\mathrm{dn}}_{lm\tilde{\omega}} = -\frac{2\pi b}{\mathrm{i}\omega}(\mathrm{i})^m\,{}_0S_l^m(\pi - \gamma; a\tilde{\omega})\delta(\omega - \tilde{\omega}), \qquad (3.26)$$

when ${}_0R^{\rm up}_{lm\omega}$ and ${}_0R^{\rm dn}_{lm\omega}$ are normalized to unit amplitude at infinity

$${}_0R^{\rm up}_{lm\omega} \underset{r\to\infty}{\simeq} \frac{e^{i\omega r^*}}{r}, \quad {}_0R^{\rm dn}_{lm\omega} \underset{r\to\infty}{\simeq} \frac{e^{-i\omega r^*}}{r}. \tag{3.27}$$

3.5 Electromagnetic plane waves

Following the prescription given in the previous sections we begin with a solution of Maxwell's equations representing a plane wave travelling up the z-axis:

$$A_0 = 0, \quad A_x = A\cos\omega(t - z), \quad A_y = A\sin\omega(t - z), \quad A_z = 0, \tag{3.28}$$

where A is a constant representing the amplitude of this left circularly polarized (positive helicity) wave at $z = t = 0$. Right circular polarization is obtained by substituting $\omega \to -\omega$. An active rotation of this solution through angle γ in the positive ('right-hand rule') sense about the x-axis followed by a transformation to spherical coordinates yields

$$\begin{aligned} &A_0 = 0, \\ &A_r = A\sin\theta(\cos\phi\cos\chi + \cos\gamma\sin\phi\sin\chi) - A\sin\gamma\cos\theta\sin\chi, \\ &A_\theta = Ar\cos\theta(\cos\phi\cos\chi + \cos\gamma\sin\phi\sin\chi) + Ar\sin\gamma\sin\theta\sin\chi, \\ &A_\phi = Ar\sin\theta(-\sin\phi\cos\chi + \cos\gamma\cos\phi\sin\chi), \end{aligned} \tag{3.29}$$

where χ is the phase defined in (3.23).

The transverse components of the vector potential are

$$A_m = A_\mu m^\mu, \quad A_{\bar{m}} = A_\mu \bar{m}^\mu = \overline{(A_m)}, \tag{3.30}$$

where m^μ and $\bar{m}^\mu$ are the complex legs of the Kinnersley (1969) tetrad (2.21). Since we need to know A_m and $A_{\bar{m}}$ only asymptotically we let $r \gg a$ (a is the angular momentum of the black hole) in (2.21) and obtain

$$\begin{aligned} A_m \cong \frac{A}{4\sqrt{2}} \{ &[e^{i(\chi-\phi)}(\cos\theta + 1) + e^{i(\phi-\chi)}(\cos\theta - 1)](1 + \cos\gamma) \\ &+ [e^{i(\chi+\phi)}(\cos\theta - 1) + e^{-i(\phi+\chi)}(\cos\theta + 1)](1 - \cos\gamma) \\ &- 2i\sin\gamma\sin\theta(e^{i\chi} - e^{-i\chi})\}; \end{aligned} \tag{3.31}$$

the complex conjugate equation holds for $A_{\bar{m}}$.

When $\gamma = 0$ the term in (3.31) that appears to have a 'helicity' of zero (no dependence on ϕ) disappears. This term arises because the rotated vector potential has three terms, only two of which are independent; the helicity zero term is without physical significance in regard to the real helicity states of the electromagnetic field. (For a discussion including the analogous phenomenon for gravitational waves, see section 3.6 below and also Weinberg (1972), p. 255).)

A_m is of spin-weight 1 and $A_{\bar{m}}$ is of spin-weight -1. See Campbell (1971) for a discussion of spin weight. Combine (3.31) and its complex conjugate into one equation:

$$\begin{aligned} {}_sA = \frac{A}{4\sqrt{2}}\{ & [\mathrm{e}^{\mathrm{i}(\chi-\phi)}(\cos\theta + s) + \mathrm{e}^{-\mathrm{i}(\chi-\phi)}(\cos\theta - s)](1+\cos\gamma) \\ & + [\mathrm{e}^{\mathrm{i}(\chi+\phi)}(\cos\theta - s) + \mathrm{e}^{-\mathrm{i}(\chi+\phi)}(\cos\theta + s)](1-\cos\gamma) \\ & - 2\mathrm{i}\sin\gamma\sin\theta(\mathrm{e}^{\mathrm{i}\chi} - \mathrm{e}^{-\mathrm{i}\chi})\} \end{aligned} \tag{3.32}$$

where

$$ {}_1A = A_m \tag{3.33}$$

and

$$ {}_{-1}A = A_{\bar{m}}. \tag{3.34}$$

Now expand ${}_sA$ in frequency and in spheroidal harmonics of the appropriate spin-weight:

$$ {}_sA = \int_{-\infty}^{\infty} \mathrm{d}\bar{\omega} \sum_{l'm'} {}_sA_{l'm'}(\bar{\omega}) {}_sZ_{l'}^{m'}(\theta, \phi; a\bar{\omega})\mathrm{e}^{-\mathrm{i}\bar{\omega}t}. \tag{3.35}$$

We will eventually require a slightly different expansion, in which the explicit parity of the modes is considered. It is simplest, however, to extract the parity dependence after developing the form (3.35). Acting on both sides of (3.35) with

$$\int_{\Omega}\int_{-\infty}^{\infty} \mathrm{d}t\, \mathrm{d}\Omega\, {}_s\bar{Z}_l^m(\theta, \phi; a\tilde{\omega})\mathrm{e}^{\mathrm{i}\tilde{\omega}t}$$

yields

$$ {}_sA_{lm}(\tilde{\omega}) = \frac{1}{2\pi}\int_{\Omega}\int_{-\infty}^{\infty} \mathrm{d}t\, \mathrm{d}\Omega\, {}_sA\mathrm{e}^{\mathrm{i}\tilde{\omega}t} {}_s\bar{Z}_l^m((\theta, \phi; a\tilde{\omega})) \tag{3.36}$$

where we have used the orthonormality of the ${}_sZ_l^m$ and the relation

$$\int_{-\infty}^{\infty} \mathrm{d}t\mathrm{e}^{\mathrm{i}(\tilde{\omega}-\bar{\omega})t} = 2\pi\delta(\tilde{\omega} - \bar{\omega}).$$

Now it is clear from (3.32) and (3.35) that ${}_sA_{lm}$ can contain only ω and $-\omega$. Drop the primes on l' and m'. From (3.23) and the definition of ${}_sA$ (3.32) we see that the integration over time in (3.36) can be performed immediately yielding

$$\begin{aligned} {}_sA_{lm} = \frac{A}{4\sqrt{2}}\{ & [{}_s\mathscr{I}_{11-1}\delta(\omega + \tilde{\omega}) + {}_s\mathscr{I}_{-1-11}\delta(\omega - \tilde{\omega})](1+\cos\gamma) \\ & + [{}_s\mathscr{I}_{1-11}\delta(\omega + \tilde{\omega}) + {}_s\mathscr{I}_{-11-1}\delta(\omega - \tilde{\omega})](1-\cos\gamma) \\ & - 2\mathrm{i}\sin\gamma[{}_s\mathscr{I}_1\delta(\omega + \tilde{\omega}) - {}_s\mathscr{I}_{-1}\delta(\omega - \tilde{\omega})]\} \end{aligned} \tag{3.37}$$

where the symbols ${}_s\mathscr{I}_{a_1}$ and ${}_s\mathscr{I}_{a_1a_2a_3}$ denote integrals

$$
\begin{aligned}
{}_s\mathscr{I}_{a_1} &= \int d\Omega \exp\{-i(a_1\rho + m\phi)\} \sin\theta\, {}_sS_l^m(\theta; a\tilde{\omega}) \\
{}_s\mathscr{I}_{a_1a_2a_3} &= \int d\Omega \exp\{-i[a_1\rho - (a_3 - m)\phi]\}(\cos\theta + a_2 s)\, {}_sS_l^m(\theta; a\tilde{\omega})
\end{aligned}
\tag{3.38a}
$$

$$s = \pm 1$$

which are evaluated in Appendix A. Here ρ is used to denote the spatial part of the exponent χ of (3.23):

$$\rho = \omega r(\sin\gamma \sin\theta \sin\phi + \cos\gamma \cos\theta).$$

The result is, after some algebra,

$$
\begin{aligned}
{}_sA_{lm}(\tilde{\omega}) = &-\frac{\pi A}{\sqrt{2i\omega r}}(s+1)[e^{-i\omega r}(-i)^{m+1}\, {}_sS_l^m(\gamma; a\tilde{\omega})\delta(\omega + \tilde{\omega}) \\
&+ e^{-i\omega r}(i)^{m+1}\, {}_sS_l^m(\pi - \gamma; a\tilde{\omega})\delta(\omega - \tilde{\omega})] \\
&+ \frac{\pi A}{\sqrt{2i\omega r}}(s-1)[e^{i\omega r}(-i)^{m+1}\, {}_sS_l^m(\gamma; a\tilde{\omega})\delta(\omega - \tilde{\omega}) \\
&+ e^{i\omega r}(i)^{m+1}\, {}_sS_l^m(\pi - \gamma; a\tilde{\omega})\delta(\omega + \tilde{\omega})].
\end{aligned}
\tag{3.39}
$$

Substitution of (3.39) into (3.35) yields (valid for $s = \pm 1$)

$$
\begin{aligned}
{}_sA = &-\frac{(s+1)}{2}\frac{\sqrt{2}A\pi}{i\omega r}\sum_{lm}[e^{-i\omega(r-t)}(-i)^{m+1}\, {}_sS_l^m(\gamma; -a\omega)\, {}_sZ_l^m(\theta, \phi; -a\omega) \\
&+ e^{-i\omega(r+t)}(i)^{m+1}\, {}_sS_l^m(\pi - \gamma; a\omega)\, {}_sZ_l^m(\theta, \phi; a\omega)] \\
&+ \frac{(s-1)}{2}\frac{\sqrt{2}A\pi}{i\omega r}\sum_{lm}[e^{i\omega(r-t)}(-i)^{m+1}\, {}_sS_l^m(\gamma; a\omega)\, {}_sZ_l^m(\theta, \phi; a\omega) \\
&+ e^{i\omega(r+t)}(i)^{m+1}\, {}_sS_l^m(\pi - \gamma; -a\omega)\, {}_sZ_l^m(\theta, \phi; -a\omega)].
\end{aligned}
\tag{3.40}
$$

Equation (3.40) is the relation we seek. It reduces in the limit $\gamma \to 0$ to equations (B2) and (B3) of CMSR, namely,

$$
\begin{aligned}
{}_1A = A_m \underset{\gamma\to 0}{\sim} \frac{\sqrt{2}A\pi}{i\omega r}\sum_l[{}_1N_{l;\pi}^1(a\omega)\, {}_1Z_l^1(\theta, \phi; a\omega)e^{-i\omega(r+t)} \\
- {}_1N_{l;0}^{-1}(-a\omega)\, {}_1Z_l^{-1}(\theta, \phi; -a\omega)e^{-i\omega(r-t)}]
\end{aligned}
\tag{3.41a}
$$

and

$$
\begin{aligned}
{}_{-1}A = A_{\bar{m}} \underset{\gamma\to 0}{\sim} \frac{\sqrt{2}A\pi}{i\omega r}\sum_l[-{}_{-1}N_{l;\pi}^{-1}(-a\omega)\, {}_{-1}Z_l^{-1}(\theta, \phi; -a\omega)e^{i\omega(r-t)} \\
+ {}_{-1}N_{l;0}^1(a\omega)\, {}_{-1}Z_l^1(\theta, \phi; a\omega)e^{i\omega(r-t)}].
\end{aligned}
\tag{3.41b}
$$

Now, insert the substitution $r \to r^*$ in the phases in (3.41) to obtain a result valid for the Kerr background. Then match the expansions for the incident plane wave above with the asymptotic limits of expansions of the ingoing (dn) and outgoing (up) portions of the vector potential in terms of the normal modes of the electromagnetic field. The normal mode expansions are given in table 2.1, page 20 (Chrzanowski, 1975); at this point we explicitly introduce a sum over parity.

First consider the ingoing piece of the vector potential A_μ^{dn}. In the ingoing gauge listed in table 2.1, A_μ^{dn} may be expanded as follows

$$A_\mu^{\mathrm{dn}} = \int_{-\infty}^{\infty} \mathrm{d}\tilde{\omega} \sum_{l,m,P} K_{lm\tilde{\omega}P}^{\mathrm{dn}} A_\mu^{\mathrm{dn}}(x, lm\tilde{\omega}P). \tag{3.42}$$

To make use of the information of table 2.1, we note that the quantities ${}_sR_{lm\omega}$ solve a homogeneous linear equation, and can be arbitrarily normalized. We will use a convenient conventional (CMSR) normalization. If we normalize the radial function in the formula for the ingoing-gauge vector potential in table 2.1 to give

$${}_{-1}R^{\mathrm{dn}}\mathrm{e}^{-\mathrm{i}\tilde{\omega}t} \sim \frac{-1}{2\mathrm{i}\tilde{\omega}} \frac{\exp[-\mathrm{i}\tilde{\omega}(r^* + t)]}{r}, \tag{3.43}$$

then asymptotically A_m^{dn} (obtained by contracting (3.42) with m^μ) behave like

$$A_m^{\mathrm{dn}} \sim \int_{-\infty}^{\infty} \mathrm{d}\tilde{\omega} \sum_{l,m,P} K_{lm\tilde{\omega}P}^{\mathrm{dn}} \, {}_1Z_l^m(\theta, \phi; a\tilde{\omega}) \frac{\exp[-\mathrm{i}\tilde{\omega}(r^* + t)]}{r}. \tag{3.44}$$

A_m is the asymptotically gauge independent part of A_μ, and so is the appropriate variable to consider for the scattering problem.

The analogous analysis to (3.42)–(3.44) may now be carried out, but contracting with $\bar{m}^\mu$, and an equation similar to (3.44) is obtained. However A_μ is real, so $A_{\bar{m}} = \overline{(A_m)}$. The equality of these two forms gives a condition on the coefficients $K_{lm\tilde{\omega}P}$

$$\bar{K}_{lm\tilde{\omega}P} = PK_{l-m-\tilde{\omega}P}(-1)^{(l+m+s)} \tag{3.45}$$

where the power of (-1) arises from (2.43)–(2.44). Equation (3.45) is a correction to similar equations in CMSR, Handler (1979), Handler & Matzner (1980), and Futterman & Matzner (1981, 1982), and is a consequence of the correction in table 2.1 as mentioned earlier in discussing (2.15).

When the expression (3.40) is compared with (3.44), and the reality condition is imposed, we have

$$K^{\mathrm{dn}}_{lm\tilde{\omega}P} = \frac{A\pi}{\mathrm{i}\omega\sqrt{2}}[(\mathrm{i})^{m+1}{}_1S^m_l(\pi-\gamma; a\tilde{\omega})\delta(\omega-\tilde{\omega}) - (\mathrm{i})^{l+m+s}P(\mathrm{i})^{m-1}{}_1S^{-m}_l(\pi-\gamma; -a\tilde{\omega})\delta(\omega+\tilde{\omega})]. \tag{3.46}$$

It may be useful to give some detail in this example. Treat the $s = +1$ part of A^{dn}. From (3.40)

$$A^{\mathrm{dn}} = -\sqrt{2}A\pi\int \mathrm{d}\tilde{\omega}\sum_{lm}\delta(\omega-\tilde{\omega})\mathrm{i}^m\frac{\mathrm{e}^{-\mathrm{i}\tilde{\omega}(t+r^*)}}{\tilde{\omega}r}{}_1S^m_l(\pi-\gamma, a\tilde{\omega})_sZ^m_l(\theta,\phi,a\tilde{\omega}).$$

Comparing this equation to (3.44) shows that

$$\delta(\omega-\tilde{\omega})\frac{\mathrm{i}^m}{\tilde{\omega}}{}_1S^m_l(\pi-\gamma, a\tilde{\omega}) = \sum_P K_{lm\tilde{\omega}P}.$$

This, together with the reality condition (3.45) gives two linear equations which can be solved for $K_{lm\tilde{\omega}1}$ and $K_{lm\tilde{\omega}-1}$, giving (3.46). The outgoing piece of A^{up} may be expanded

$$A^{\mathrm{up}}_\mu = \int_{-\infty}^{\infty}\mathrm{d}\tilde{\omega}\sum_{l,m,P}K^{\mathrm{up}}_{lm\tilde{\omega}P}A^{\mathrm{up}}_\mu(x, lm\tilde{\omega}P). \tag{3.47}$$

In the outgoing gauge of table 2.1, the asymptotic form of $A^{\mathrm{up}}_{\bar{m}}$ is

$$A^{\mathrm{up}}_{\bar{m}} = \int_{-\infty}^{\infty}\mathrm{d}\tilde{\omega}\sum_{l,m,P}K^{\mathrm{up}}_{lm\tilde{\omega}P}\,{}_{-1}Z^m_l(\theta, a\tilde{\omega})\left(\frac{\exp[\mathrm{i}\tilde{\omega}(r^*-t)]}{r}\right), \tag{3.48}$$

if (normalization)

$${}_1R^{\mathrm{up}}\mathrm{e}^{-\mathrm{i}\tilde{\omega}t}\underset{r\to\infty}{\sim}\frac{1}{\mathrm{i}\tilde{\omega}}\frac{\exp[\mathrm{i}\tilde{\omega}(r^*-t)]}{r^3}. \tag{3.49}$$

When the outgoing part of (3.40) is matched to (3.48) we find, using (2.43) and (3.45), that

$$K^{\mathrm{up}}_{lm\omega P} = (-1)^{l+1}K^{\mathrm{dn}}_{lm\omega P}. \tag{3.50}$$

The standard normalizations of the remaining radial functions are:

$${}_1R^{\mathrm{dn}}\mathrm{e}^{-\mathrm{i}\omega t}\sim 4\mathrm{i}\omega\frac{\exp[-\mathrm{i}\omega(r^*+t)]}{r} \tag{3.51}$$

$${}_{-1}R^{\mathrm{up}}\mathrm{e}^{-\mathrm{i}\omega t}\sim\frac{-2\mathrm{i}\omega r}{B^2}\exp[\mathrm{i}\omega(r^*-t)]. \tag{3.52}$$

The constant B in (3.52) is required for the normalizations of ${}_1R^{\mathrm{up}}$ and

${}_{-1}R^{\mathrm{up}}$ to be consistent with the Teukolsky–Starobinsky identities (see Chandrasekhar (1979b) for details), and is given by (2.59).

Equation (3.50) is a correction to equation (36) of Futterman & Matzner (1981) and equation (B11) of CMSR, in which the factor $(-1)^{l+m}$ appears instead of $(-1)^{l+1}$. Since these papers discuss axially incident radiation only, the only values of m that occur are $m = \pm 1$. Thus no other results in these papers are affected.

The K^{dn}, K^{up} of (3.46) and (3.50) are natural generalizations of those of the axial incidence case of CMSR. For axial incidence,

$$ {}_1S^{\pm m}(\pi; a\omega) = {}_1N^{\pm 1}_{l;\pi}(a\omega)\delta_{1,\pm m} \tag{3.53}$$

where ${}_1N^{\pm 1}_{l;\pi}$ is defined by (2.48). Equations (3.46) and (3.50) thus reduce to the on axis result of CMSR. K^{up} and K^{dn} completely specify a plane wave solution of Maxwell's equations.

With the aid of table 2.1, we obtain formulae for the electromagnetic field:

$$\begin{aligned}
-2\phi_0^{\mathrm{dn}} &= \int_{-\infty}^{\infty} \mathrm{d}\tilde{\omega} \sum_{l,m,P} K^{\mathrm{dn}}_{lm\tilde{\omega}P}\, {}_1R^{\mathrm{dn}}\, {}_1Z_l^m(a\tilde{\omega})\mathrm{e}^{-\mathrm{i}\tilde{\omega}t},\\
-2\rho^{-2}\phi_2^{\mathrm{dn}} &= \int_{-\infty}^{\infty} \mathrm{d}\tilde{\omega} \sum_{l,m,P} BK^{\mathrm{dn}}_{lm\tilde{\omega}P}\, {}_{-1}R^{\mathrm{dn}}\, {}_{-1}Z_l^m(a\tilde{\omega})\mathrm{e}^{-\mathrm{i}\tilde{\omega}t},\\
-2\phi_0^{\mathrm{up}} &= \int_{-\infty}^{\infty} \mathrm{d}\tilde{\omega} \sum_{l,m,P} BK^{\mathrm{up}}_{lm\tilde{\omega}P}\, {}_1R^{\mathrm{up}}\, {}_1Z_l^m(a\tilde{\omega})\mathrm{e}^{-\mathrm{i}\tilde{\omega}t},\\
-2\rho^{-2}\phi_2^{\mathrm{up}} &= \int_{-\infty}^{\infty} \mathrm{d}\tilde{\omega} \sum_{l,m,P} B^2K^{\mathrm{up}}_{lm\tilde{\omega}P}\, {}_{-1}R^{\mathrm{up}}\, {}_{-1}Z_l^m(a\tilde{\omega})\mathrm{e}^{-\mathrm{i}\tilde{\omega}t}.
\end{aligned} \tag{3.54}$$

The relative normalizations of ϕ_0 and ϕ_2 are consistent with the Teukolsky–Starobinsky identities mentioned above. Asymptotic values for the plane-wave fields $\phi_{0,2}^{\mathrm{up,dn}}$ can be obtained by taking the appropriate limits of (3.54).

An alternative method, based on the field tensor, may be used to derive (3.54). We consider only axial incidence in this example. Start with the formulae for ϕ_0 and ϕ_2 for a flat-space plane wave

$$\begin{aligned}
\phi_0 &= \frac{\mathrm{i}\omega A}{\sqrt{2}}(1-\cos\theta)\exp[\mathrm{i}\omega(z-t)],\\
\phi_2 &= \frac{\mathrm{i}\omega A}{2\sqrt{2}}(1+\cos\theta)\exp[\mathrm{i}\omega(z-t)].
\end{aligned} \tag{3.55}$$

Then the expansion of (3.55) in spheroidal harmonics

$$\phi_0 \sim \sum_l -2\sqrt{2\pi\omega}A\left({}_1N^1_{l;\pi}(a\omega)\frac{\exp[-i\omega(r+t)]}{\omega r}\,{}_1Z^1_l(\theta, a\omega)\right.$$
$$\left. +2\,{}_1N^1_{l;0}(a\omega)\frac{\exp i\omega(r-t)}{(\omega r)^3}\,{}_1Z^1_l(\theta, a\omega)\right) \tag{3.56}$$

and

$$\phi_2 \sim \sum_l \sqrt{2\pi\omega}A\left({}_{-1}N^1_{l;0}(a\omega)\frac{\exp[i\omega(r-t)]}{(\omega r)}\,{}_{-1}Z^1_l(\theta, a\omega)\right.$$
$$\left. +2\,{}_{-1}N^1_{l;\pi}(a\omega)\frac{\exp[-i\omega(r+t)]}{(\omega r)^3}\,{}_{-1}Z^1_l(\theta, a\omega)\right) \tag{3.57}$$

gives the correct asymptotic forms of ϕ_0 and ϕ_2 when the substitution $r \to r^*$ is made in the exponents. Notice that the ingoing and outgoing pieces of ϕ_0 and ϕ_2 have radial amplitude dependences that differ by r^2. This is an example of the 'peeling' property which generally gives the ratio of powers as $r^{|2s|}$ (NP); see also section 2.4.

3.6 Gravitational plane waves

We begin with a left circularly polarized transverse traceless wave travelling up the z-axis of a rectangular Minkowski coordinate system in flat spacetime. The metric perturbation is

$$h_{\mu\nu} = h\begin{bmatrix} 0 & 0 & 0 & 0 \\ 0 & \cos\chi & \sin\chi & 0 \\ 0 & \sin\chi & -\cos\chi & 0 \\ 0 & 0 & 0 & 0 \end{bmatrix} \tag{3.58}$$

where $\chi = \omega(t-z)$.

An active rotation of (3.58) through angle γ counterclockwise about the positive x-axis yields

$$h_{\mu\nu} = h\begin{bmatrix} 0 & 0 & 0 & 0 \\ 0 & \cos\chi & \cos\gamma\sin\chi & -\sin\gamma\sin\chi \\ 0 & \cos\gamma\sin\chi & -\cos^2\gamma\cos\chi & \cos\gamma\sin\gamma\cos\chi \\ 0 & -\sin\gamma\sin\chi & \cos\gamma\sin\gamma\cos\chi & -\sin^2\gamma\cos\chi \end{bmatrix} \tag{3.59}$$

with χ now defined as in (3.23), and where γ is the angle between the incident propagation vector and the positive z-axis. We now transform (3.59) to spherical coordinates which are the asymptotic limit of spheroidal coordinates. The results are listed in Appendix B; a typical term is

$$\begin{aligned} h_{rr} = h_{11} = h[&\sin^2\theta(\cos^2\phi - \sin^2\phi\cos^2\gamma) - \cos^2\theta\sin^2\gamma \\ &+ 2\sin\theta\cos\theta\sin\phi\sin\gamma\cos\gamma]\cos\chi \\ &+ 2h(\sin^2\theta\cos\phi\sin\phi\cos\gamma - \sin\theta\cos\theta\cos\phi\sin\gamma)\sin\chi. \end{aligned} \tag{3.60}$$

Compute

$$\begin{aligned} h_{mm} &= h_{\mu\nu}m^\mu m^\nu \\ h_{\bar{m}\bar{m}} &= h_{\mu\nu}\bar{m}^\mu\bar{m}^\nu = \overline{(h_{mm})}. \end{aligned} \tag{3.61}$$

Observe that h_{mm} is of spin-weight 2 and $h_{\bar{m}\bar{m}}$ is of spin-weight -2, and write

$$\begin{aligned} {}_{+2}h &= h_{mm}, \\ {}_{-2}h &= h_{\bar{m}\bar{m}}. \end{aligned} \tag{3.62}$$

Instead of the form given in (2.21) for m^μ, take the asymptotic limit of this form to obtain the asymptotic limit:

$${}_sh = \frac{h_{22}}{2r^2} + \left(\frac{s}{2}\right)\frac{ih_{23}}{r^2\sin\theta} - \frac{h_{33}}{2r^2\sin^2\theta}, \tag{3.63}$$

whence

$$\begin{aligned} {}_sh = (h/16)\{&[-6\sin^2\gamma\sin^2\theta](e^{i\chi} + e^{-i\chi}) \\ &- 4i(\cos\gamma - 1)\sin\gamma\sin\theta[(\tfrac{1}{2}s - \cos\theta)e^{i(\chi+\phi)} + (\tfrac{1}{2}s + \cos\theta)e^{-i(\chi+\phi)}] \\ &- 4i(\cos\gamma + 1)\sin\gamma\sin\theta[(\tfrac{1}{2}s + \cos\theta)e^{i(\chi-\phi)} + (\tfrac{1}{2}s - \cos\theta)e^{-i(\chi-\phi)}] \\ &+ (\cos\gamma - 1)^2[(\tfrac{1}{2}s - \cos\theta)^2e^{i(\chi+2\phi)} + (\tfrac{1}{2}s + \cos\theta)^2e^{-i(\chi+2\phi)}] \\ &+ (\cos\gamma + 1)^2[(\tfrac{1}{2}s + \cos\theta)^2e^{i(\chi-2\phi)} + (\tfrac{1}{2}s - \cos\theta)^2e^{-i(\chi-2\phi)}]\}. \end{aligned} \tag{3.64}$$

We now expand (3.64) in spin-weighted spheroidal harmonics

$${}_sh = \int_{-\infty}^{\infty} d\tilde{\omega} \sum_{lm} {}_s\mathscr{B}_{lm}(\tilde{\omega})\,{}_sZ_l^m(\theta, \phi; a\tilde{\omega})e^{-i\tilde{\omega}t}. \tag{3.65}$$

The argument here is similar to that in the previous section. Inverting (3.65) we obtain

$${}_s\mathscr{B}_{lm}(\tilde{\omega}) = \frac{1}{2\pi}\int_{-\infty}^{\infty}\int_{\Omega} dt\, d\Omega\, {}_sh\, {}_s\bar{Z}_l^m(\theta, \phi; a\tilde{\omega})e^{i\tilde{\omega}t}. \tag{3.66}$$

Inserting (3.64) in (3.66) and performing the integration over time yields

$$\begin{aligned} {}_s\mathscr{B}_{lm}(\omega) = (h/16)\{&-6\sin^2\gamma[{}_s\mathscr{I}_1^0\delta(\omega + \tilde{\omega}) + {}_s\mathscr{I}_{-1}^0\delta(\omega - \tilde{\omega})] \\ &- 4i(\cos\gamma - 1)\sin\gamma[{}_s\mathscr{I}_{1-11}^1\delta(\omega + \tilde{\omega}) + {}_s\mathscr{I}_{-11-1}^1\delta(\omega - \tilde{\omega})] \\ &- 4i(\cos\gamma + 1)\sin\gamma[{}_s\mathscr{I}_{11-1}^1\delta(\omega + \tilde{\omega}) + {}_s\mathscr{I}_{-1-11}^1\delta(\omega - \tilde{\omega})] \\ &+ (\cos\gamma - 1)^2[{}_s\mathscr{I}_{11-1}^2\delta(\omega + \tilde{\omega}) + {}_s\mathscr{I}_{-11-1}^2\delta(\omega - \tilde{\omega})] \\ &+ (\cos\gamma + 1)^2[{}_s\mathscr{I}_{11-1}^2\delta(\omega + \tilde{\omega}) + {}_s\mathscr{I}_{-1-11}^2\delta(\omega - \tilde{\omega})]\} \end{aligned} \tag{3.67}$$

where the integrals $\mathscr{I}$ are defined by

$$ {}_s\mathscr{I}^0_{a_1} = \int \mathrm{d}\Omega \exp[-\mathrm{i}(a_1\rho + m\phi)] \sin^2\theta\,{}_sS_l^m(\theta; a\tilde{\omega}) \tag{3.68a} $$

$$ {}_s\mathscr{I}^1_{a_1a_2a_3} = \int \mathrm{d}\Omega \exp\{-\mathrm{i}[a_1\rho + (m - a_3)\phi]\} \times \sin\theta(s/2 + a_2\cos\theta)\,{}_sS_l^m(\theta; a\tilde{\omega}) \tag{3.68b} $$

$$ {}_s\mathscr{I}^2_{a_1a_2a_3} = \int \mathrm{d}\Omega \exp\{-\mathrm{i}[a_1\rho + (m - 2a_3)\phi]\} \times (s/2 + a_2\cos\theta)^2\,{}_sS_l^m(\theta; a\tilde{\omega}) \tag{3.68c} $$

and a_1, a_2, $a_3 = \pm 1$.

The result is (see Appendix A)

$$ \begin{aligned} {}_s\mathscr{B}_{lm}(\tilde{\omega}; \gamma) = {} & \frac{2\pi h}{\mathrm{i}\omega r}\left(\frac{s+2}{4}\right)[(-\mathrm{i})^m \mathrm{e}^{-\mathrm{i}\omega r}\,{}_sS_l^m(\gamma; a\tilde{\omega})\delta(\omega + \tilde{\omega}) \\ & + (\mathrm{i})^m \mathrm{e}^{-\mathrm{i}\omega r}\,{}_sS_l^m(\pi - \gamma; a\tilde{\omega})\delta(\omega - \tilde{\omega})] \\ & + \frac{2\pi h}{\mathrm{i}\omega r}\left(\frac{s-2}{4}\right)[(-\mathrm{i})^m \mathrm{e}^{\mathrm{i}\omega r}\,{}_sS_l^m(\gamma; a\tilde{\omega})\delta(\omega - \tilde{\omega}) \\ & + (\mathrm{i})^m \mathrm{e}^{\mathrm{i}\omega r}\,{}_sS_l^m(\pi - \gamma; a\tilde{\omega})\delta(\omega + \tilde{\omega})] \end{aligned} \tag{3.69} $$

where it is understood that $s = \pm 2$.

Substituting (3.69) into (3.65) and performing the integration over $\tilde{\omega}$ yields

$$ \begin{aligned} {}_sh = {} & \frac{2\pi h}{\mathrm{i}\omega r}\left(\frac{s+2}{4}\right)\sum_{lm}[(-\mathrm{i})^m \mathrm{e}^{-\mathrm{i}\omega(r-t)}\,{}_sS_l^m(\gamma; -a\omega)\,{}_sZ_l^m(\theta, \phi; -a\omega) \\ & + (\mathrm{i})^m \mathrm{e}^{-\mathrm{i}\omega(r+t)}\,{}_sS_l^m(\pi - \gamma; a\omega)\,{}_sZ_l^m(\theta, \phi; a\omega)] \\ & + \frac{2\pi h}{\mathrm{i}\omega r}\left(\frac{s-2}{4}\right)\sum_{lm}[(-\mathrm{i})^m \mathrm{e}^{\mathrm{i}\omega(r-t)}\,{}_sS_l^m(\gamma; a\omega)\,{}_sZ_l^m(\theta, \phi; a\omega) \\ & + (\mathrm{i})^m \mathrm{e}^{\mathrm{i}\omega(r+t)}\,{}_sS_l^m(\pi - \gamma; -a\omega)\,{}_sZ_l^m(\theta, \phi; -a\omega)], \end{aligned} \tag{3.70} $$

which reduces in the limit $\gamma \to 0$ to equations (3.4) and (3.5) of CMSR, namely:

$$ {}_2h = h_{mm} \underset{\gamma\to 0}{\sim} -\frac{2\pi h}{\mathrm{i}\omega r}\sum_l [{}_2N_{l:0}^{-2}(-a\omega)\,{}_2Z_l^{-2}(\theta, \phi; -a\omega)\mathrm{e}^{-\mathrm{i}\omega(rt)} + {}_2N^2{}_{l:\pi}(a\omega)\,{}_2Z_l^2(\theta, \phi; a\omega)\mathrm{e}^{-\mathrm{i}\omega(r+t)}] \tag{3.71a} $$

$$ {}_{-2}h = h_{\bar{m}\bar{m}} \underset{\gamma\to 0}{\sim} \frac{2\pi h}{\mathrm{i}\omega r}\sum_l [{}_{-2}N_{l:0}^2(a\omega)\,{}_{-2}Z_l^2(\theta, \phi; a\omega)\mathrm{e}^{\mathrm{i}\omega(r-t)} + {}_{-2}N_{l:\pi}^{-2}(-a\omega)\,{}_{-2}Z_l^{-2}(\theta, \phi; -a\omega)\mathrm{e}^{\mathrm{i}\omega(r+t)}]. \tag{3.71b} $$

The expansions

$$h_{mm} = \sum {}_2\mathscr{B}_{lm\,2}Z_l^m(\theta, \phi; a\omega) \tag{3.72}$$

$$h_{\bar{m}\bar{m}} = \sum {}_{-2}\mathscr{B}_{lm\,-2}Z_l^m(\theta, \phi; a\omega) \tag{3.73}$$

must be matched to the asymptotic forms of the normal mode expansions for the same quantities. We write

$$h_{\mu\nu}^{\mathrm{dn}} = \int_{-\infty}^{\infty} \mathrm{d}\tilde{\omega} \sum_{lmP} K_{lm\tilde{\omega}P}^{\mathrm{dn}} h_{\mu\nu}^{\mathrm{dn}}(lm\tilde{\omega}P) \tag{3.74}$$

where $h_{\mu\nu}(lm\tilde{\omega}P)$ is the $lm\tilde{\omega}P$ mode of the perturbing metric as given in table 2.1 (in the ingoing gauge). The asymptotic form of the transverse tracefree part is

$$h_{mm}^{\mathrm{dn}} \simeq \int_{-\infty}^{\infty} \mathrm{d}\tilde{\omega} \sum_{lmP} K_{lm\tilde{\omega}P\,2}^{\mathrm{dn}} Z_l^m(\theta, \phi; a\tilde{\omega}) \frac{\exp[-\mathrm{i}\tilde{\omega}(t + r^*)]}{r} \tag{3.75}$$

when one takes the normalization

$$_{-2}R^{\mathrm{dn}}\mathrm{e}^{-\mathrm{i}\tilde{\omega}t} \simeq \frac{1}{4\tilde{\omega}^2}\exp[-\mathrm{i}\tilde{\omega}(t + r^*)]. \tag{3.76}$$

Because the metric is real we again have the 'crossing relation'

$$\bar{K}_{lm\tilde{\omega}P} = PK_{l-m-\tilde{\omega}P}(-1)^{l+m+s}. \tag{3.77}$$

Comparison of (3.75) with the ingoing part of (3.72) using (3.77) (and again replacing $r \to r^*$ in exponents) yields

$$K_{lm\tilde{\omega}P}^{\mathrm{dn}} = \left(\frac{\pi h}{\mathrm{i}\omega}\right)[(\mathrm{i})^m{}_2S_l^m(\pi - \gamma; a\tilde{\omega})\delta(\omega - \tilde{\omega})$$
$$-(-1)^{l+m+s}P(\mathrm{i})^m{}_2S_l^{-m}(\pi - \gamma; -a\tilde{\omega})\delta(\omega + \tilde{\omega})]. \tag{3.78}$$

Similarly,

$$h_{\mu\nu}^{\mathrm{up}} = \int_{-\infty}^{\infty} \mathrm{d}\tilde{\omega} \sum_{lmP} K_{lm\tilde{\omega}P}^{\mathrm{up}} h_{\mu\nu}^{\mathrm{up}}(lm\tilde{\omega}P) \tag{3.79}$$

where $h_{\mu\nu}^{\mathrm{up}}(lm\tilde{\omega}P)$ is the $lm\tilde{\omega}P$ mode of the metric perturbation in the outgoing gauge in table 2.1, so that

$$h_{\bar{m}\bar{m}}^{\mathrm{up}} \simeq \int_{-\infty}^{\infty} K_{lm\tilde{\omega}P\,-2}^{\mathrm{up}} Z_l^m(\theta, \phi; a\tilde{\omega}) \frac{\exp[-\mathrm{i}\tilde{\omega}(t - r^*)]}{r} \tag{3.80}$$

with the normalization (CMSR):

$$_2R^{\mathrm{up}}\mathrm{e}^{-\mathrm{i}\tilde{\omega}t} = \frac{1}{\tilde{\omega}^2}\frac{\exp[-\mathrm{i}\tilde{\omega}(t - r^*)]}{r^5}. \tag{3.81}$$

Comparison of (3.80) with the outgoing part of (3.70) using (3.77) yields

$$K^{\text{up}}_{lm\tilde{\omega}P} = (-1)^{l+1} K^{\text{dn}}_{lm\tilde{\omega}P}. \tag{3.82}$$

Equation (3.82) represents a correction of equation (3.15) of CMSR. That paper deals with axially incident perturbations only; therefore, no other results of CMSR are affected.

With the aid of Table 2.1 and the properties of the Teukolsky functions we obtain formulas for the NP field quantities valid at all radii. These formulae, given in CMSR, are

$$
\begin{aligned}
-8\Psi_0^{\text{dn}} &= \int_{-\infty}^{\infty} \mathrm{d}\tilde{\omega} \sum_{lmP} K^{\text{dn}}_{lm\tilde{\omega}P\,2}R^{\text{dn}}_{lm\tilde{\omega}\,2}Z^m_l(\theta,\phi;a\tilde{\omega})\mathrm{e}^{-\mathrm{i}\tilde{\omega}t},\\
-8\rho^{-4}\Psi_4^{\text{dn}} &= \int_{-\infty}^{\infty} \mathrm{d}\tilde{\omega} \sum_{lmP} (\mathrm{Re}\,C\\
&\quad + 12\mathrm{i}M\tilde{\omega}P)K^{\text{dn}}_{lm\tilde{\omega}P\,-2}R^{\text{dn}}_{lm\tilde{\omega}\,-2}Z^m_l(\theta,\phi;a\tilde{\omega})\mathrm{e}^{-\mathrm{i}\tilde{\omega}t},\\
-8\Psi_0^{\text{up}} &= \int_{-\infty}^{\infty} \mathrm{d}\tilde{\omega} \sum_{lmP} (\mathrm{Re}\,C\\
&\quad - 12\mathrm{i}M\tilde{\omega}P)K^{\text{up}}_{lm\tilde{\omega}P\,2}R^{\text{up}}_{lm\tilde{\omega}\,2}Z^m_l(\theta,\phi;a\tilde{\omega})\mathrm{e}^{-\mathrm{i}\tilde{\omega}t},\\
-8\rho^{-4}\Psi_4^{\text{up}} &= \int_{-\infty}^{\infty} \mathrm{d}\tilde{\omega} \sum_{lmP} |C|^2 K^{\text{up}}_{lm\tilde{\omega}P\,-2}R^{\text{up}}_{lm\tilde{\omega}\,-2}Z^m_l(\theta,\phi;a\tilde{\omega})\mathrm{e}^{-\mathrm{i}\tilde{\omega}t},
\end{aligned} \tag{3.83}
$$

where

$$C = \mathrm{Re}\,C + 12\mathrm{i}M\tilde{\omega}; \tag{3.84}$$

see (2.61)–(2.62).

The conventional CMSR normalizations of the remaining radial functions are

$${}_2R^{\text{dn}}\mathrm{e}^{-\mathrm{i}\tilde{\omega}t} \simeq 16\tilde{\omega}^2 \frac{\exp[-\mathrm{i}\tilde{\omega}(t+r^*)]}{r} \tag{3.85}$$

and

$${}_{-2}R^{\text{up}}\mathrm{e}^{-\mathrm{i}\tilde{\omega}t} \simeq \frac{4r^3\tilde{\omega}^2}{|C|^2}\exp[-\mathrm{i}\tilde{\omega}(t-r^*)]. \tag{3.86}$$

Using the normalizations (3.76), (3.81), (3.85) and (3.86) will enable us to determine asymptotic forms for the Weyl tensor components $\Psi_0^{\substack{\text{up}\\\text{dn}}}$ and $\Psi_4^{\substack{\text{up}\\\text{dn}}}$

The normalizations for the radial functions given above are consistent with the large r forms of the Teukolsky–Starobinsky identities, and the constant C is derived from them. See Chandrasekhar (1979b, 1983) and references cited therein for a readable discussion.

The Riemann tensor components Ψ_0 and Ψ_4 for a distorted plane wave may be derived by more direct alternate means that do not necessitate the introduction of formulae for the Kerr metric perturbations. This second method for deriving formulae for Ψ_0 and Ψ_4 entails making the substitution $r \to r^*$ in the flat-space formulae for Ψ_0^{down} and Ψ_4^{up} and using the NP equations to determine the r^{-5} terms Ψ_0^{up} and Ψ_4^{down}. As in the electromagnetic case, we give only the on-axis case for the field-quantity perturbation.

For a plane gravitational wave travelling up the z axis on a flat background, it is straightforward to show that

$$\Psi_0 = b(1-\cos\theta)^2 \mathrm{e}^{2\mathrm{i}\phi} \exp[\mathrm{i}\omega(z-t)], \tag{3.87}$$

$$\Psi_4 = \tfrac{1}{4} b(1+\cos\theta)^2 \mathrm{e}^{2\mathrm{i}\phi} \exp[\mathrm{i}\omega(z-t)], \tag{3.88}$$

($b = -\frac{1}{2}h\omega^2$, where h is the amplitude of the metric fluctuation), which may be expanded in terms of spheroidal harmonics to find the asymptotic forms

$$\Psi_0 \sim \sum_l 8\pi \frac{b}{\mathrm{i}} \left(\frac{24\,{}_2N^2_{l;0}(a\omega)}{(\omega r)^5} \mathrm{e}^{\mathrm{i}\omega r} - \frac{{}_2N^2_{l;\pi}(a\omega)\mathrm{e}^{-\mathrm{i}\omega r}}{\omega r} \right) {}_2Z^2_l(a\omega)\mathrm{e}^{-\mathrm{i}\omega t} \tag{3.89}$$

and

$$\Psi_4 = \sum_l 2\pi \frac{b}{\mathrm{i}} \left({}_{-2}N^2_{l;0}(a\omega) \frac{\mathrm{e}^{\mathrm{i}\omega r}}{\omega r} - \frac{24\,{}_{-2}N^2_{l;\pi}(a\omega)\mathrm{e}^{-\mathrm{i}\omega r}}{(\omega r)^5} \right) {}_{-2}Z^2_l(a\omega)\mathrm{e}^{-\mathrm{i}\omega t}. \tag{3.90}$$

Notice the peeling property here gives asymptotic behavior separated by *four* powers of r.

We now show that if we follow the prescription of setting $r \to r^*$ in the exponents and taking the resulting expressions as the asymptotic values of ψ_0 and ψ_4 on the Kerr background, then the power flux per l mode of ingoing and outgoing radiation at infinity (as computed from ψ_4) does not balance. Using (3.90) and the Teukolsky & Press (1974) formulae for power flux in terms of ψ_4, we obtain

$$\frac{\mathrm{d}E_{\mathrm{in}}}{\mathrm{d}t} = \frac{128\omega^6}{|C|^2} \frac{4\pi^2 b^2}{\omega^{10}} (24)^2 ({}_{-2}N^2_{l;\pi})^2 \tag{3.91}$$

and

$$\frac{\mathrm{d}E_{\mathrm{out}}}{\mathrm{d}t} = \frac{1}{2\omega^2} \frac{4\pi^2 b^2}{\omega^2} ({}_{-2}N^2_{l;0})^2. \tag{3.92}$$

As shown by CMSR, their ratio is given by

$$\frac{\mathrm{d}E_{\text{out}}}{\mathrm{d}E_{\text{in}}} = \frac{({}_{-2}N^2_{l;0})^2}{256}|C|^2 \frac{1}{(24)^2}({}_{-2}N^2_{l;\pi})^{-2} \tag{3.93}$$

$$= \frac{|C|^2}{|\mathrm{Re}\, C|^2} \tag{3.94}$$

and is unity only if $M = 0$, i.e., on a flat background. A similar discrepancy arises when the Ψ_0 ingoing and outgoing power-flux formulae are compared.

However, if we use the r^{-1} part of Ψ_0 to describe the ingoing flux and the r^{-1} part of Ψ_4 to describe the outgoing flux, then we find their ratio to be unity. Therefore it must be concluded that when $M \neq 0$ the coefficients of the r^{-5} terms are incorrect as given by (3.89) and (3.90).

To correct the r^{-5} pieces in the expressions for Ψ_0 and Ψ_4, we must use the Bianchi identities together with the presumably correct r^{-1} terms Ψ_4^{up} and Ψ_0^{down}. Equations (3.89) and (3.90) lead to the (asymptotic) identification

$$\Psi_0^{\text{down}} \sim \sum_l -\frac{8\pi b}{\mathrm{i}}\, {}_2N^2_{l;\pi} \frac{\exp[-\mathrm{i}\omega(r^* + t)]}{\omega r}\, {}_2Z_l^2(a\omega) \tag{3.95}$$

and

$$\Psi_4^{\text{up}} \sim \sum_l \frac{2\pi b}{\mathrm{i}}\, {}_{-2}N^2_{l;0} \frac{\exp[\mathrm{i}\omega(r^* - t)]}{\omega r}\, {}_{-2}Z_l^2(a\omega). \tag{3.96}$$

Since Ψ_0 and Ψ_4 completely determine each other (Wald, 1973), we may find Ψ_0^{up} from Ψ_4^{up} and Ψ_0^{down} from Ψ_0^{down}. Specifically, the Bianchi identities together with the spin-coefficient equations lead to the connection formulae

$$\begin{aligned} \mathrm{D}^4\rho^{-4}\Psi_4 = {} & \rho^{-4}(\bar{\delta} + 3\alpha + \bar{\beta})(\bar{\delta} + 2\alpha + 2\bar{\beta})(\bar{\delta} + \alpha + 3\bar{\beta})(\bar{\delta} + 4\bar{\beta})\Psi_0 \\ & - 3M(\partial/\partial t)\bar{\Psi}_0 \end{aligned} \tag{3.97}$$

and

$$\begin{aligned} & \rho^{-4}(\Delta + 5\bar{\mu} + 3\gamma - \bar{\gamma})(\Delta + 5\bar{\mu} + 2\gamma - 2\bar{\gamma})(\Delta + 5\bar{\mu} + \gamma - 3\bar{\gamma})(\Delta + \bar{\mu} - 4\bar{\gamma})\Psi_0 \\ & = (\delta + 5\bar{\pi} - 3\beta - \bar{\alpha})(\delta + 5\bar{\pi} - 2\beta - 2\bar{\alpha})(\delta + 5\bar{\pi} - \beta - 3\bar{\alpha})(\delta + \bar{\pi} \\ & \quad - 4\bar{\alpha})\rho^{-4}\Psi_4 + 3M(\partial/\partial t)\bar{\Psi}_4. \end{aligned} \tag{3.98}$$

Notice that no separation into parity states has been made in (3.89), (3.90), nor correspondingly in (3.95), (3.96). Substitution of the 'down' and 'up' formulae into, respectively, (3.97) and (3.98) then gives (3.83) summed over parity states.

The result summed over parities has a remarkable feature: Only one value of m appears in the ingoing (outgoing) part of Ψ_0 (Ψ_4), while both m and its negative appear in the outgoing (ingoing) parts. For instance, consider the composite (summed over parity) form of Ψ_4:

$$\Psi_0 \sim \sum_l \frac{4\pi h\omega^2}{i} {}_2N^2_{l;\pi}(a\omega)\left[\frac{\exp[-i\omega(r^*+t)]}{\omega r} {}_2S^2_l(a\omega)e^{2i\phi}\right.$$
$$-(-1)^{l+m}\left(\frac{\mathrm{Re}\, C}{16}\frac{\exp[i\omega(r^*-t)]}{\omega^5 r^5} {}_2S^2_l(a\omega)e^{2i\phi}\right.$$
$$\left.\left.-\frac{3iM\omega}{4}\frac{\exp[-i\omega(r^*-t)]}{\omega^5 r^5} {}_2S^{-2}_l(-a\omega)e^{-2i\phi}\right)\right]. \qquad (3.99)$$

An $m = -2$ term occurs in the outgoing piece only. When we consider the sum over parity, this anomalous outgoing piece has no convenient interpretation nor do the connection formulae (3.97), (3.98) used to derive (3.99). In addition, this extra term makes the summed-over parity-states result unsuitable for the purpose of numerical calculations.

To make sense out of the connection formulae and the resultant expressions for the distorted plane wave, one must consider separately the scattering of each parity state. Notice from (3.78) and (3.82) that each parity solution has both ingoing and outgoing parts for each m; connection formulae (3.97) and (3.98) relate in a comprehensible fashion Ψ_0 and Ψ_4 for the ingoing or outgoing piece of either parity state. It is an unfortunate interference between the two parity solutions which renders the peculiar form of (3.99).

The appearance of a nontrivial parity dependence in the field-tensor expression for the fiducial plane wave may be at first surprising since such a dependence does not arise in the expression for the electromagnetic field variables ϕ_0, ϕ_2 in (3.56) and (3.57). Furthermore the separated Teukolsky equations make no reference to parity. The parity dependence of the reference plane wave will guarantee that gravitational wave scattering is parity dependent, despite this parity independence of the Teukolsky equation. This is consistent with the behavior in the Schwarzschild case for metric perturbations, where Zerilli (1970) and Moncrief (1975) have shown that the *equations* obeyed by the metric perturbations do depend on parity. We have just seen that the *plane-wave* metric perturbations themselves depend only trivially on the parity. The two approaches (metric as opposed to Riemann tensor) put the parity dependence into different aspects of the problem. In this context working with the Riemann perturbations Ψ_0, Ψ_4 may be simpler because the field equation need only be solved once; the parity dependence appears only algebraically in the normalization arising from the plane wave. Chandrasekhar & Detweiler (1975) have shown the relation between the two parity solutions of the metric perturbation equations. They demonstrate that the difference is only one of *phase*, consistent with the way parity appears as $\operatorname{Re} C - 12iM\omega P$ in (3.83).

The Riemann–tensor parity differences exist because C is complex. Investigation of the Bianchi identities in which C arises shows that $\operatorname{Im} C$ is attributable to the fact that one is perturbing a field (Ψ_2) whose background value is nonzero ($\Psi_2 = M\rho^3$), and this is thus also the reason for the parity dependence of the metric perturbation equations. For an analysis of this problem based on metric perturbations, see Chandrasekhar (1978b).

3.7 Aside: incident plane waves in the Reissner–Nordstrøm spacetime

The Reissner (1916)–Nordstrøm (1918) spacetime respects the gravitational and electromagnetic field outside a charged black hole. The metric is

$$\begin{aligned} ds^2 = &-\left(1 - \frac{2M}{r} + \frac{q^2}{r^2}\right) dt^2 \\ &+ \left(1 - \frac{2M}{r} + \frac{q^2}{r^2}\right)^{-1} dr^2 + r^2(d\theta^2 + \sin^2\theta \, d\phi^2), \end{aligned} \tag{3.100}$$

where M is the mass and q is the charge of the black hole. The only nonvanishing component of the electromagnetic field is

$$F_{or} = \frac{q}{r^2}\left(1 - \frac{2M}{r} + \frac{q^2}{r^2}\right)^{-1/2}. \tag{3.101}$$

In this case, because there is a background electromagnetic field, electromagnetic waves combine with the background to produce stress-tensor perturbations which are first order in the perturbing field, rather than second order as is typically the case. Then first-order electromagnetic disturbances are coupled to first-order gravitational ones. This can lead to interconversion between electric and gravitational wave modes, and in fact the natural modes of the system will be a combination of gravitational and electromagnetic waves. In this case one can simply consider an incident plane wave which is a combination of the on-axis version of the electromagnetic waves ((3.42), (3.46) with (3.50)) and the on-axis version of a gravitational plane wave ((3.74), (3.77), (3.78)). (The on-axis case is appropriate because of the spherical symmetry.) Chandrasekhar (1979a) has given an extensive discussion of Reissner–Nordstrøm perturbations.

The first-order electromagnetic fluctuation makes a first-order fluctuation appear in the metric. We can consistently consider both of them (or a linear combination) to be propagating on the fixed background. Moncrief

(1975) gives the explicit formulae expressing the independent modes $R_\pm$ in terms of the electromagnetic and gravitational fields. To be consistent with Moncrief's notation we will use $\pm$ to denote the two combination decoupled modes; we will use the terms even, odd to stand for $P = +1, P = -1$, respectively.

In Moncrief's formulation $R_\pm^{\text{odd,even}}$ satisfy

$$\left(\frac{\mathrm{d}^2}{\mathrm{d}r^{*2}} + \omega^2 - V_\pm^{\text{odd}}\right) R_\pm^{\text{odd}} = 0, \tag{3.102}$$

$$\left(\frac{\mathrm{d}^2}{\mathrm{d}r^{*2}} + \omega^2 - V_\pm^{\text{even}}\right) R_\pm^{\text{even}} = 0; \tag{3.103}$$

the even and odd modes are decoupled from one another. The frequency dependence in Fourier modes is everywhere taken as $\mathrm{e}^{-i\omega t}$. Here the potentials $V_\pm^{\text{odd,even}}$ are

$$V_\pm^{\text{odd}} = N\mathrm{e}^{-\lambda}\left(\frac{l(l+1)}{r^2} - \frac{3M}{r^3} + \frac{4q^2}{r^4} \pm \frac{\sigma}{r^3}\right); \tag{3.104}$$

and

$$V_\pm^{\text{even}} = N\mathrm{e}^{-\lambda}(V \pm \sigma S), \tag{3.105}$$

with

$$V = \frac{N\mathrm{e}^{-\lambda}}{(r\Lambda)^2}\left(\frac{8q^2}{r^2} - \frac{6M}{r}\right)^2 + \frac{8q^2 N\mathrm{e}^{-\lambda}}{r^4\Lambda} + \frac{l(l+1)(l-1)(l+2)}{r^2\Lambda}$$
$$+ \frac{3M}{r^3} + \frac{4q^2}{r^4\Lambda}\left(2 - \frac{6M}{r} + \frac{4q^2}{r^2}\right) \tag{3.106}$$

and

$$S = \frac{l(l+1)}{r^3\Lambda} + \frac{2N\mathrm{e}^{-\lambda}}{r^3\Lambda^2}\left((l-1)(l+2) + \frac{4q^2}{r^2}\right) - \frac{1}{r^3\Lambda}\left(\frac{2M}{r} - \frac{2q^2}{r^2}\right). \tag{3.107}$$

In (3.102)–(3.107) we used the Reissner–Nordstrøm definitions:

$$\mathrm{d}r/\mathrm{d}r^* = N\mathrm{e}^{-\lambda} = 1 - 2M/r + q^2/r^2, \tag{3.108}$$

$$\Lambda = (l-1)(l+2) + (6M/r) - (4q^2/r^2), \tag{3.109}$$

and

$$\sigma = [9M^2 + 4q^2(l-1)(l+2)]^{1/2}. \tag{3.110}$$

In both the even- and odd-parity cases the variables $R_\pm$ are linear combinations of electromagnetic and gravitational radiation. The gravitational incident perturbation then will consist, for example, of

$$Q^{\text{odd,even}} = \sin\psi R_+^{\text{odd,even}} + \cos\psi R_-^{\text{odd,even}} \tag{3.111}$$

where the angle ψ is defined by

$$\sin 2\psi = -2Pq[(l-1)(l+2)]^{1/2}/\sigma. \tag{3.112}$$

The electromagnetic perturbation will be

$$H^{\text{odd,even}} = \cos\psi R_{+}^{\text{odd,even}} - \sin\psi R_{-}^{\text{odd,even}}. \tag{3.113}$$

Notice that in both the odd- and even-parity cases the difference between the equations for $R_{\pm}$ is of the order σr^{-3} while the dominant terms in the equations are of the order $\omega^2 - l(l+1)/r^2$. Hence the $R_{\pm}$ behave similarly near infinity, and, near infinity one may specify a particular mixture of electromagnetic and gravitational incident radiation. Nearer the black hole, however, (3.102) and (3.103) show that the electromagnetic-gravitational mixture will change since R_{+} and R_{-} evolve differently.

The metric variables h_{mm}^{down}, $h_{\bar{m}\bar{m}}^{\text{up}}$ (describing the wave in appropriate asymptotically transverse traceless gauges) are asymptotically related to Q by

$$h_{mm(\text{odd})}^{\text{down}} \underset{r\to\infty}{\sim} \mathrm{i}\frac{[l(l+1)]^{1/2}}{[(l-1)(l+2)]^{1/2}}\frac{{}_2Y_l^m}{r}\mathrm{e}^{-\mathrm{i}\omega t}Q_{(\text{odd})}^{\text{down}}, \tag{3.114}$$

$$h_{\bar{m}\bar{m}}^{\text{up(odd)}} \underset{r\to\infty}{\sim} \mathrm{i}\frac{[l(l+1)]^{1/2}}{[(l-1)(l+2)]^{1/2}}\frac{{}_{-2}Y_l^m}{r}\mathrm{e}^{-\mathrm{i}\omega t}Q_{(\text{odd})}^{\text{up}}, \tag{3.115}$$

$$h_{mm}^{\text{down(even)}} \underset{r\to\infty}{\sim} \frac{Q_{(\text{even})}^{\text{down}}}{[l(l+1)]^{1/2}}\frac{{}_2Y_l^m}{r}\mathrm{e}^{-\mathrm{i}\omega t}, \tag{3.116}$$

$$h_{\bar{m}\bar{m}}^{\text{up(even)}} \underset{r\to\infty}{\sim} \frac{Q_{(\text{even})}^{\text{up}}}{[l(l+1)]^{1/2}}\frac{{}_{-2}Y_l^m}{r}e^{-\mathrm{i}\omega t}. \tag{3.117}$$

or, using the expansions (3.74) and (3.79)

$$Q_{(\text{odd})}^{\text{down}} \sim \frac{1}{\mathrm{i}}\frac{[(l-1)(l+2)]^{1/2}}{[l(l+1)]^{1/2}}K_{l,m,\tilde{\omega},P=-1}^{\text{down}}\mathrm{e}^{-\mathrm{i}\tilde{\omega}r^*}, \tag{3.118}$$

$$Q_{(\text{odd})}^{\text{up}} \sim \frac{1}{\mathrm{i}}\frac{[(l-1)(l+2)]^{1/2}}{[l(l+1)]^{1/2}}K_{l,m,\tilde{\omega},P=-1}^{\text{up}}\mathrm{e}^{\mathrm{i}\tilde{\omega}r^*}, \tag{3.119}$$

$$Q_{(\text{even})}^{\text{down}} \sim [l(l+1)]^{1/2}K_{l,m,\tilde{\omega},P=+1}^{\text{down}}\mathrm{e}^{-\mathrm{i}\tilde{\omega}r^*}, \tag{3.120}$$

$$Q_{(\text{even})}^{\text{up}} \sim [l(l+1)]^{1/2}K_{l,m,\tilde{\omega},P=+1}^{\text{up}}\mathrm{e}^{\mathrm{i}\tilde{\omega}r^*}. \tag{3.121}$$

At infinity, the radiation field associated with the electromagnetic perturbation (3.113) is contained in the transverse components of $\mathbf{E}$: $E_{\hat{\theta}}$ and $E_{\hat{\phi}}$. For even parity,

$$E_m = E_\alpha m^\alpha = \frac{1}{\sqrt{2}}(E_{\hat{\theta}} + E_{\hat{\phi}}) \sim \frac{H_{,r}^{\text{even}}\,{}_1Y_l^m}{2\sqrt{2}r[l(l+1)]^{1/2}}. \tag{3.122}$$

Also

$$E_{\bar{m}} \sim \frac{-H^{\text{even}}_{,r}\ {}_{-1}Y_l^m}{2\sqrt{2}r[l(l+1)]^{1/2}}. \tag{3.123}$$

Clearly the quantity $2|E_m|^2$ gives the sum of the squares of the transverse components in an orthonormal frame. For the odd case, the transverse component of the vector potential (which is invariant under odd-parity gauge transformations) is

$$A_m = \frac{-\mathrm{i}}{\sqrt{2}r}\frac{[l(l+1)]^{1/2}}{[(l-1)(l+2)]^{1/2}}H^{\text{odd}}\ {}_1Y_l^m. \tag{3.124}$$

The two variables E_m, A_m, if of the same parity, are related by

$$E_m = A_{m,t} \tag{3.125}$$

in a transverse-traceless gauge. From (3.111)–(3.125) and the asymptotic behavior of $R_\pm$,

$$H^{\text{up}}_{(\text{even})} \sim -2\sqrt{2}[l(l+1)]^{1/2}K^{\text{up}}_{l,m,\bar{\omega},P=+1}\mathrm{e}^{\mathrm{i}\bar{\omega}r^*}, \tag{3.126}$$

$$H^{\text{down}}_{(\text{even})} \sim -2\sqrt{2}[l(l+1)]^{1/2}K^{\text{down}}_{l,m,\bar{\omega},P=+1}\mathrm{e}^{-\mathrm{i}\bar{\omega}r^*}, \tag{3.127}$$

$$H^{\text{up}}_{(\text{odd})} \sim \frac{-2\sqrt{2}[(l-1)(l+2)]^{1/2}}{[l(l+1)]^{1/2}}K^{\text{up}}_{l,m,\bar{\omega},P=-1}\mathrm{e}^{\mathrm{i}\bar{\omega}r^*}, \tag{3.128}$$

$$H^{\text{down}}_{(\text{odd})} \sim \frac{-2\sqrt{2}}{\mathrm{i}}\frac{[(l-1)(l+1)]^{1/2}}{[l(l+1)]^{1/2}}K^{\text{down}}_{l,m,\bar{\omega},P=-1}\mathrm{e}^{-\mathrm{i}\bar{\omega}r^*}, \tag{3.129}$$

where the $K^{\text{up,down}}_{lm\bar{\omega}P}$ are those quantities defined in (3.46), (3.50), (3.78) and (3.82) for electromagnetic and for gravitational plane waves respectively.

The discussion of the Reissner–Nordstrøm case has followed that of Matzner (1976). The general Reissner–Nordstrøm perturbations have been discussed by Chandrasekhar (1979a).

4
Neutrino plane waves

4.1 Introduction

There is no potential for the neutrino field, which means that one must work directly with the field quantities which have a more complicated asymptotic power-law behavior in $(1/r)$ than either metric perturbations or the vector potential. The peeling theorem (cf section 2.4) predicts that we will have to deal with asymptotic solutions differing by *one* power of r at infinity.

This complicates the integrations necessary to perform the mode-expansions. In addition the neutrino fields transform under changes of coordinates and tetrads in a more complicated way than do vector or tensor quantities. Both of these features are due to the neutrino's intrinsic spin-$\frac{1}{2}$ character.

The interaction of neutrinos and gravitational fields was first studied by Brill & Wheeler (1957), who investigated several aspects of that problem including the bound states of neutrinos in a spherically symmetric gravitational field.

More recently neutrinos in the Kerr background have been studied by Unruh (1973), Teukolsky (1973), Chandrasekhar (1976) and Chandrasekhar & Detweiler (1977). The results of these investigations are summarized in Chandrasekhar (1979b; 1983). Briefly, the two-component neutrino and Dirac equations have been shown to be separable in the Kerr geometry, and it has been shown that unlike integer spin fields, neutrinos and electrons do not exhibit classical superradiance in the Kerr background.

In this chapter we expand neutrino plane waves in the normal modes appropriate to the Kerr geometry. We give an elementary account of electron and neutrino plane waves in the NP formalism in flat spacetime. We then transform the flat spacetime plane waves to a tetrad and coordinate system appropriate to the asymptotic Kerr geometry, expand in spin-$\frac{1}{2}$ spheroidal harmonics, and match to normal mode expansions.

4.2 Neutrino waves in flat background spacetimes

As a prelude to our treatment of neutrinos, we consider first the Dirac equation. In natural units ($\hbar = 1$) it takes the form:

$$(\gamma^\mu \nabla_\mu + i m_e)\psi = 0 \tag{4.1}$$

where ∇_μ is a spinorial covariant derivative (see Rosenman (1971) for a detailed discussion, also NP and (4.17) and (4.18) below). We use a representation in which

$$\gamma^\mu = \sqrt{2}\begin{bmatrix} 0 & (\sigma^{\mu A\dot{B}})^T \\ (\sigma^\mu{}_{A\dot{B}}) & 0 \end{bmatrix} \tag{4.2}$$

and

$$\psi = \begin{bmatrix} P^A \\ Q_{\dot{B}} \end{bmatrix}. \tag{4.3}$$

The symbols $\sigma^\mu_{A\dot{B}}$ are the Infeld – van der Waerden symbols so mercilessly suppressed by Penrose (1968). For each value of μ (spacetime labels are denoted by Greek letters) $\sigma^\mu{}_{A\dot{B}}$ is a 2×2 matrix. We choose them to be Hermitian. The upper case Latin labels are abstract spinor indices, with a dot over an index indicating that that index transforms by the complex conjugated transformation. The objects P^A and $Q_{\dot{B}}$ are 2-component spinors.

The condition that iteration of (4.1) yield the Klein–Gordon equation becomes

$$\sigma^\mu_{A\dot{B}}\sigma^\nu_{C\dot{D}}g_{\mu\nu} = \varepsilon_{AC}\varepsilon_{\dot{B}\dot{D}} \tag{4.4}$$

where $\varepsilon_{AC} = \varepsilon_{\dot{A}\dot{C}} = \varepsilon^{AC} = \varepsilon^{\dot{A}\dot{C}}$ is a 2×2 antisymmetric spinor

$$\varepsilon_{AC} = \begin{pmatrix} 0 & 1 \\ -1 & 0 \end{pmatrix}. \tag{4.5}$$

(equation (4.4) holds whether $g_{\mu\nu}$ is flat or not.)

Since the $\sigma^\mu_{A\dot{B}}$ are considered to be maps from tensors to spinors via the correspondence

$$V_{A\dot{B}} \equiv \sigma^\mu{}_{A\dot{B}} V_\mu \tag{4.6}$$

where V_μ is any 4-vector, the symbols ε_{AC} are called the spin metric and are used to raise and lower spinor indices:

$$P^A \varepsilon_{AC} = P_C$$
$$\varepsilon^{AC} P_C = P^A. \tag{4.7}$$

The order is important because of the antisymmetry of ε_{AC}. In fact

$$P^A Q_A = -P_A Q^A \tag{4.8}$$

so that for any 2-spinor

$$P^A P_A = -P_A P^A = 0. \tag{4.9}$$

Equation (4.1) may be rewritten in terms of 2-spinors:

$$\nabla_{A\dot{B}} P^A + \mathrm{i}\mu_e Q_{\dot{B}} = 0 \tag{4.10a}$$

$$\nabla_{B\dot{A}} \bar{Q}^B + \mathrm{i}\mu_e \bar{P}_{\dot{A}} = 0 \tag{4.10b}$$

where

$$\sqrt{2}\mu_e = m_e \tag{4.11a}$$

and

$$\nabla_{A\dot{B}} \equiv \sigma^{\mu}_{A\dot{B}} \nabla_{\mu}. \tag{4.11b}$$

Equation (4.10b) requires that the $\sigma^{\mu}{}_{A\dot{B}}$ be Hermitian. Thus far we have considered quantities in abstract spin space without any reference to a particular basis. We have no way of assigning values to the components of $\sigma^{\mu}_{A\dot{B}}$ and thus no way of assigning values to the components of the γ-matrices. DeWitt (1964) has shown that spinors have meaning in a curved spacetime only when they are defined in a locally flat region about each point of the spacetime, i.e., when they are defined with respect to an orthonormal, or a null, tetrad field.

Let ι^A and o^A be basis spinors with $\iota^A o_A = 1$ and define

$$\begin{aligned}
\sigma^{\mu}{}_{A\dot{B}} \iota^A \bar{\iota}^{\dot{B}} &= l^{\mu}, \\
\sigma^{\mu}{}_{A\dot{B}} \iota^A \bar{o}^{\dot{B}} &= m^{\mu}, \\
\sigma^{\mu}{}_{A\dot{B}} o^A \bar{\iota}^{\dot{B}} &= \bar{m}^{\mu}. \\
\sigma^{\mu}{}_{A\dot{B}} o^A \bar{o}^{\dot{B}} &= n^{\mu}.
\end{aligned} \tag{4.12}$$

It is easily shown using (4.4) that the vectors $l^{\mu}, m^{\mu}, \bar{m}^{\mu}, n^{\mu}$ are null; moreover, they have the same inner products as the vectors of a NP null tetrad

$$l^{\mu} n_{\mu} = -m^{\mu} \bar{m}_{\mu} = 1 \tag{4.13}$$

and all other inner products are zero.

We define a spinor dyad

$$\rho^A{}_0 = \iota^A \quad \rho^A{}_1 = o^A \tag{4.14}$$

and write

$$\sigma^{\mu}{}_{ab} = \sigma^{\mu}{}_{A\dot{B}} \rho^A{}_a \bar{\rho}^{\dot{B}}{}_{\dot{b}} \tag{4.15}$$

where lower case Latin indices denote dyad components. In this dyad,

$$\sigma^{\mu}_{0\dot{0}} = l^{\mu} \quad \sigma^{\mu}_{0\dot{1}} = m^{\mu},$$
$$\sigma^{\mu}_{1\dot{0}} = \bar{m}^{\mu}, \quad \sigma^{\mu}_{1\dot{1}} = n^{\mu}, \tag{4.16}$$

give explicit formulae for the components of $\sigma^{\mu}{}_{a\dot{b}}$ in terms of a null tetrad.

4.3 Neutrino waves in curved spacetimes

Thus far, the spatial curvature has not entered our discussion. However, in general, we must use the spinor *covariant* derivative which is chosen so that $\sigma^{\mu}{}_{A\dot{B}}$ and $\varepsilon_{AC}\varepsilon_{\dot{B}\dot{D}}$ are covariantly constant since they are in effect the metric tensor $g_{\mu\nu}$ with one and two tensor indices replaced by spinor indices. The spinor covariant derivative must also be invariant under general coordinate transformations and tetrad rotations (Lorentz spinor transformations). Finally, it must satisfy the usual properties of ordinary covariant differentiation: the Leibniz property, linearity, equivalence to partial differentiation when applied to a scalar, and reduction to partial differentiation in Minkowski spacetime. The spinor covariant derivative which satisfies these properties was given by NP:

$$\nabla_{\mu}\mathscr{E}^{A} = \mathscr{E}^{A}{}_{,\mu} + \Gamma^{A}{}_{D\mu}\mathscr{E}^{D}$$
$$\nabla_{\mu}\mathscr{E}_{B} = \mathscr{E}_{B,\mu} - \mathscr{E}_{D}\Gamma^{D}{}_{B\mu}. \tag{4.17}$$

Here $\mathscr{E}^{A}$ is an arbitrary spinor and $\Gamma^{A}_{B\mu}$ is the spinor connection given by

$$\Gamma^{A}_{B\mu} = \tfrac{1}{2}\sigma^{A\dot{C}}_{\alpha}(\sigma^{\alpha}_{B\dot{C},\mu} + \sigma^{\beta}_{B\dot{C}}\Gamma^{\alpha}_{\beta\mu}) \tag{4.18}$$

where $\Gamma^{\alpha}_{\beta\mu}$ is the usual affine connection.

The covariant derivative of a spinor may be expressed in terms of its components along the dyad basis $\rho^{A}{}_{a}$ as

$$\rho^{A}{}_{a}\bar{\rho}^{\dot{B}}{}_{\dot{b}}\rho^{c}{}_{C}\nabla_{A\dot{B}}\mathscr{E}^{C} = \partial_{a\dot{b}}\mathscr{E}^{c} + \Gamma^{c}_{da\dot{b}}\mathscr{E}^{d} \tag{4.19}$$

where

$$\partial_{a\dot{b}} = \sigma^{\mu}_{a\dot{b}}\partial_{\mu}$$

are directional derivatives along the tetrad legs. Standard notation is

$$\partial_{0\dot{0}} = \mathrm{D} = l^{\mu}\partial_{\mu}, \quad \partial_{1\dot{0}} = \bar{\delta} = m^{\mu}\partial_{\mu},$$
$$\partial_{0\dot{1}} = \delta = m^{\mu}\partial_{\mu}, \quad \partial_{1\dot{1}} = \Delta = n^{\mu}\partial_{\mu}. \tag{4.20}$$

Writing out the first of (4.10) in dyad components, we obtain the pair of equations

$$(\mathrm{D} + \Gamma_{100\dot{0}} - \Gamma_{001\dot{0}})P^{0} + (\bar{\delta} + \Gamma_{110\dot{0}} - \Gamma_{011\dot{0}})P^{1} = \mathrm{i}\mu_{e}Q^{\dot{1}},$$
$$(\delta + \Gamma_{100\dot{1}} - \Gamma_{001\dot{1}})P^{0} + (\Delta + \Gamma_{110\dot{1}} - \Gamma_{011\dot{1}})P^{1} = -\mathrm{i}\mu_{e}Q^{\dot{0}}. \tag{4.21}$$

Table 4.1. *Notation for spin coefficients*

	cd \ ab	00	01 or 10	11
$\Gamma_{abc\dot{d}} =$	$0\dot{0}$	κ	ε	π
	$1\dot{0}$	ρ	α	λ
	$0\dot{1}$	σ	β	μ
	$1\dot{1}$	τ	γ	ν

From Newman & Penrose (1962).

A similar pair of equations is provided by the second of equations (4.10). The symbols $\Gamma_{abc\dot{d}}$ are the spin coefficients of NP; a standard notation for these spin coefficients is given in table 4.1; cf chapter 2. The NP equations for the Dirac field are thus

$$\begin{aligned}
&(\mathrm{D}-\varepsilon-\rho)P^0+(\bar{\delta}+\pi-\alpha)P^1=\mathrm{i}\mu_e Q^{\dot{1}},\\
&(\Delta+\mu-\gamma)P^1+(\delta+\beta-\tau)P^0=-\mathrm{i}\mu_e Q^{\dot{0}}\\
&(\mathrm{D}-\bar{\varepsilon}-\bar{\rho})Q^{\dot{0}}+(\delta+\bar{\pi}-\bar{\alpha})Q^{\dot{1}}=-\mathrm{i}\mu_e P^1,\\
&(\Delta+\bar{\mu}-\bar{\gamma})Q^{\dot{1}}+(\bar{\delta}+\bar{\beta}-\bar{\tau})Q^{\dot{0}}=\mathrm{i}\mu_e P^0.
\end{aligned}\tag{4.22}$$

4.4 Neutrino waves in Kerr background; separation of variables

The spin coefficients for the Kinnersley (1969) tetrad in Boyer–Lindquist (1967) coordinates for the Kerr metric are known: they are listed for example by Teukolsky (1973, equation (4.5)) and given in chapter 2, (2.22).

Chandrasekhar (1976) showed that equations (4.22) can be solved in this basis by separation of variables:

$$\begin{aligned}
P^0&=(r-\mathrm{i}a\cos\theta)^{-1}{}_{-1/2}R(r)_{-1/2}S(\theta)\mathrm{e}^{\mathrm{i}(m\phi-\omega t)},\\
P^1&={}_{1/2}R(r)_{1/2}S(\theta)\mathrm{e}^{\mathrm{i}(m\phi-\omega t)},\\
Q^{\dot{0}}&=-(r+\mathrm{i}a\cos\theta)^{-1}{}_{-1/2}R(r)_{1/2}S(\theta)\mathrm{e}^{\mathrm{i}(m\phi-\omega t)},\\
Q^{\dot{1}}&={}_{1/2}R(r)_{-1/2}S(\theta)\mathrm{e}^{\mathrm{i}(m\phi-\omega t)}
\end{aligned}\tag{4.23}$$

The angular and radial equations obtained by the above substitutions are

$$\begin{aligned}
{}_{1/2}\mathscr{L}_{1/2}S&=-(\not\lambda^{1/2}-am_e\cos\theta)_{-1/2}S,\\
{}_{1/2}\mathscr{L}^{\dagger}{}_{-1/2}S&=(\not\lambda^{1/2}+am_e\cos\theta)_{1/2}S,
\end{aligned}\tag{4.24}$$

and

$$\begin{aligned}
{}_0\mathscr{D}_{-1/2}R&=(\not\lambda^{1/2}+\mathrm{i}m_e r)_{1/2}R,\\
\Delta_{1/2}\mathscr{D}^{\dagger}{}_{1/2}R&=(\not\lambda^{1/2}-\mathrm{i}m_e r)_{-1/2}R,
\end{aligned}\tag{4.25}$$

where $\not\lambda$ is a separation constant and

$$
\begin{aligned}
{}_n\mathscr{L} &= \partial_\theta + (-a\omega\sin\theta + m\,\mathrm{cosec}\,\theta) + n\cot\theta,\\
{}_n\mathscr{L}^\dagger &= \partial_\theta - (-a\omega\sin\theta + m\,\mathrm{cosec}\,\theta) + n\cot\theta,
\end{aligned} \tag{4.26}
$$

$$
\begin{aligned}
{}_0\mathscr{D} &= \partial_r - \mathrm{i}K/\Delta,\\
{}_{1/2}\mathscr{D}^\dagger &= \partial_r + (r - M + \mathrm{i}K)/\Delta.
\end{aligned} \tag{4.27}
$$

(The quantity K is defined in (2.29).) The operators $\mathscr{L}$ defined in (4.26) are closely related to the $\eth$ and $\bar{\eth}$ operators of Geroch *et al.* (1973); cf. Chandrasekhar (1976, 1977).

Elimination of first ${}_{1/2}R$ and then ${}_{-1/2}R$ from (4.25) gives a pair of equations that may be expressed as a single equation with $s = \pm\frac{1}{2}$ as a parameter

$$
\Delta^{-s}\frac{\mathrm{d}}{\mathrm{d}r}\left(\Delta^{s+1}\frac{\mathrm{d}_sR}{\mathrm{d}r}\right) + \left(\frac{K^2 - 2\mathrm{i}s(r-M)K}{\Delta} + 4\mathrm{i}s\omega r - \not\lambda\right){}_sR
$$
$$
-\left[\left(\frac{2\mathrm{i}sm_\mathrm{e}\Delta}{\not p - 2\mathrm{i}sm_\mathrm{e}r}\right)\left(\frac{\mathrm{d}}{\mathrm{d}r} - \frac{2\mathrm{i}sK}{\Delta} + \frac{(2s+1)(r-M)}{2\Delta}\right) - m_\mathrm{e}^2r^2\right]{}_sR = 0, \tag{4.28}
$$

where

$$
\begin{aligned}
\not p &= (\not\lambda)^{1/2} & s &= -\tfrac{1}{2};\\
\not p &= (\not\lambda + 1)^{1/2}, & s &= +\tfrac{1}{2}.
\end{aligned} \tag{4.29}
$$

A similar procedure for the angular equations gives

$$
\frac{1}{\sin\theta}\frac{\mathrm{d}}{\mathrm{d}\theta}\left(\sin\theta\frac{\mathrm{d}}{\mathrm{d}\theta}S\right) + \left(a^2\omega^2\cos^2\theta - \frac{m^2}{\sin^2\theta} - 2a\omega s\cos\theta\right.
$$
$$
\left. - \frac{2ms\cos\theta}{\sin^2\theta} - s^2\cot^2\theta + \not\lambda - a^2\omega^2 + 2am\omega + s\right)S
$$
$$
+\left(\frac{am_\mathrm{e}\sin\theta}{\not p - 2am_\mathrm{e}s\cos\theta}\right)\left(\frac{\mathrm{d}}{\mathrm{d}\theta} - 2a\omega s\sin\theta + \frac{2ms}{\sin\theta} - a^2m_\mathrm{e}{}^2\cos^2\theta\right)S = 0. \tag{4.30}
$$

If the mass $m_\mathrm{e} = 0$, equations (4.28) and (4.30) reduce to equations derived by Unruh (1973) and Teukolsky (1973) for the massless neutrino. In particular, the $m_\mathrm{e} = 0$ equations are of the same form as Teukolsky's (1973) master equations (2.27) and (2.28).

The integration of (4.28) for the case $m_\mathrm{e} \neq 0$ is in principle no more difficult than for the massless case, once the separation constant $\not\lambda$ is given. However, for $am_\mathrm{e} \neq 0$ the angular equation becomes considerably more complicated, because $\not\lambda$ ceases to be simply an eigenvalue of a self-adjoint differential operator. The scattering of electrons in the Kerr geometry will thus be left largely for future work, although some of the machinery needed for the problem will be developed below. In particular, we continue with a 4-component formalism so that as much of the present work as possible will be applicable to electrons.

4.5 Properties of neutrino plane waves

Consider the NP form of the Dirac equation expressed again in flat spacetime. Results from this analysis will be used to determine the form of plane waves to be matched asymptotically to the normal mode expansions of spin-$\frac{1}{2}$ fields in the Kerr metric.

In the tetrad

$$\sigma^{\mu}_{0\dot{0}} = l^{\mu} = 1/\sqrt{2}(1,0,0,1),$$
$$\sigma^{\mu}_{0\dot{1}} = m^{\mu} = 1/\sqrt{2}(0,1,\mathrm{i},0) = (\overline{\sigma^{\mu}_{1\dot{0}}}),$$
$$\sigma^{\mu}_{1\dot{1}} = n^{\mu} = 1/\sqrt{2}(1,0,0,-1), \tag{4.31}$$

in flat spacetime in Cartesian coordinates the van der Waerden matrices become

$$\sigma^{0}_{a\dot{b}} = \frac{1}{\sqrt{2}}\begin{pmatrix} 1 & 0 \\ 0 & 1 \end{pmatrix}, \qquad \sigma^{1}_{a\dot{b}} = \frac{1}{\sqrt{2}}\begin{pmatrix} 0 & 1 \\ 1 & 0 \end{pmatrix},$$
$$\sigma^{2}_{a\dot{b}} = \frac{1}{\sqrt{2}}\begin{pmatrix} 0 & \mathrm{i} \\ -\mathrm{i} & 0 \end{pmatrix}, \qquad \sigma^{3}_{a\dot{b}} = \frac{1}{\sqrt{2}}\begin{pmatrix} 1 & 0 \\ 0 & -1 \end{pmatrix}, \tag{4.32}$$

which are just the $(1/\sqrt{2})$ times the usual Pauli spin matrices. Similarly the matrices with raised indices are

$$(\sigma^{ja\dot{b}})^{\mathrm{T}} = -\frac{1}{\sqrt{2}}\sigma^{j}, \quad (\sigma^{0a\dot{b}})^{\mathrm{T}} = \frac{1}{\sqrt{2}}\sigma^{0}, \tag{4.33}$$

where σ^j $(j = 1, 2, 3)$ is a Pauli matrix $(\sigma^1 = \sigma^x, \text{etc})$ and σ^0 is the identity matrix.

In what follows we use Latin indices from the beginning of the alphabet for spinor dyad components. Latin indices from the middle of the alphabet denote spatial tensor components and range from 1 to 3. Greek indices will denote 4-vector (tensor) components.

According to (4.2), (4.32) and (4.33) the γ-matrices are

$$\gamma^{j} = \begin{pmatrix} 0 & -\sigma^{j} \\ \sigma^{j} & 0 \end{pmatrix}, \quad \gamma^{0} = \begin{pmatrix} 0 & I \\ I & 0 \end{pmatrix}; \tag{4.34}$$

and the Dirac equation becomes

$$\left[\begin{pmatrix} 0 & I \\ I & 0 \end{pmatrix}\partial_0 + \overline{\begin{pmatrix} 0 & -\sigma^j \\ \sigma^j & 0 \end{pmatrix}}\partial_j + \mathrm{i}m_e\right]\begin{pmatrix} P^a \\ Q_b \end{pmatrix}. \tag{4.35}$$

The reason for the bar, which denotes complex conjugation, is that in order to use matrix multiplication with P^a and Q_b as column matrices both the σ^μ_{ab} and $(\sigma^{\mu ab})^T$ appearing in (4.2) must be transposed when used in (4.10) and (4.35). Since the σ^μ are Hermitian, transposition is equivalent to complex conjugation.

Following standard treatments, such as that given in Merzbacher (1971) or Sakurai (1967) we introduce the spin operator

$$\Sigma^i = \frac{\mathrm{i}}{2}(\gamma^j\gamma^k - \gamma^k\gamma^j)$$
$$= \overline{\begin{pmatrix} \sigma^i & 0 \\ 0 & \sigma^i \end{pmatrix}} \quad (i, j, k \text{ cyclic}), \tag{4.36}$$

the Hamiltonian

$$H = -\mathrm{i}\gamma^0\gamma^k\partial_k + \gamma^0 m_e \tag{4.37}$$

or

$$H = -\mathrm{i}\overline{\begin{pmatrix} \sigma^k & 0 \\ 0 & -\sigma^k \end{pmatrix}}\partial_k + m_e\begin{pmatrix} 0 & I \\ I & 0 \end{pmatrix}, \tag{4.38}$$

and the matrix

$$\gamma^5 = \mathrm{i}\gamma^0\gamma^1\gamma^2\gamma^3$$
$$= \begin{pmatrix} I & 0 \\ 0 & -I \end{pmatrix}. \tag{4.39}$$

The helicity operator is

$$h = \Sigma^i k_i/|k| \tag{4.40}$$

where $k_i \equiv -\mathrm{i}\partial_i$ and $k = (k_i k^i)^{1/2}$ (i ranges from 1 to 3). Using (4.36) we may write the helicity as

$$h = \frac{\bar{\sigma}^i k_i}{|k|}\begin{pmatrix} I & 0 \\ 0 & I \end{pmatrix}. \tag{4.41}$$

With the above operators we may determine, in the basis given by (4.31), solutions to the Dirac equation representing plane waves that are simultaneous eigenstates of H and h, energy and helicity.

The plane wave states are assumed to be of the form

$$\psi = \begin{pmatrix} P^a \\ Q_{\dot b} \end{pmatrix} = \begin{pmatrix} p^a \\ q_{\dot b} \end{pmatrix} \exp\{-\mathrm{i}[\omega t - k(\sin\gamma y + \cos\gamma z)]\}, \tag{4.42}$$

and p^a and $q_{\dot b}$ are constant 2-component spinors. Use of (4.38) and (4.42) in

$$\mathrm{i}\partial_t \psi = H\psi \tag{4.43}$$

yields a pair of equations

$$q_{\dot b} = \left(\frac{\omega - \bar\sigma\cdot k}{m_e}\right) p^a,$$

($\dot b = a$; i.e. they have the same numerical value)

$$p^a = \left(\frac{\omega + \bar\sigma\cdot k}{m_e}\right) q_{\dot b} \tag{4.44}$$

which are consistent only if

$$\omega^2 = k^2 + m_e{}^2, \tag{4.45}$$

the standard energy relation. We may choose either of the constant 2-spinors p^a or $q_{\dot b}$ arbitrarily; we take them to be eigenspinors $\eta_\pm$ of $\bar\sigma\cdot k/|k|$ belonging to eigenvalues $\not e = \pm 1$. Equations (4.44) now become

$$\begin{aligned} m_e q_{\dot b} &= (\omega - \not e k) p^a, \\ m_e p^a &= (\omega + \not e k) q_{\dot b}, \end{aligned} \tag{4.46}$$

from which we see that there are four independent basis 4-spinors that satisfy the Dirac equation with the coordinate dependence chosen in (4.42). These are given below

$$\omega > 0, \quad \not e = 1, \qquad\qquad \omega > 0, \quad \not e = -1,$$

$$\Lambda\begin{pmatrix} \eta_+ \\ \dfrac{\omega - k}{m_e}\,\eta_+ \end{pmatrix}, \qquad \Lambda\begin{pmatrix} \dfrac{\omega - k}{m_e}\,\eta_- \\ \eta_- \end{pmatrix}, \tag{4.47a}$$

$$\omega < 0, \quad \not e = 1, \qquad\qquad \omega < 0, \quad \not e = -1,$$

$$\Lambda\begin{pmatrix} \dfrac{\omega + k}{m_e}\,\eta_+ \\ \eta_+ \end{pmatrix}, \qquad \Lambda\begin{pmatrix} \eta_- \\ \dfrac{\omega + k}{m_e}\,\eta_- \end{pmatrix}. \tag{4.47b}$$

Here Λ is a normalization factor. If we assume the directional dependence of (4.42), the eigenspinors $\eta_{\pm}$ of $(\bar{\sigma}^i k_i)/|k|$ solve

$$\not{e}\begin{pmatrix}\eta^0\\ \eta_1\end{pmatrix}=\frac{k}{|k|}\begin{pmatrix}\cos\gamma & -\mathrm{i}\sin\gamma\\ \mathrm{i}\sin\gamma & -\cos\gamma\end{pmatrix}\begin{pmatrix}\eta^0\\ \eta_1\end{pmatrix} \tag{4.48}$$

which implies that eigen 2-spinors $\eta_{\pm}$ are given by

$$k>0, \qquad\qquad k<0,$$

$$\eta_+=\eta\begin{pmatrix}1\\ \dfrac{\mathrm{i}\sin\gamma}{\cos\gamma+1}\end{pmatrix},\qquad \eta_+=\eta\begin{pmatrix}\dfrac{-\mathrm{i}\sin\gamma}{\cos\gamma+1}\\ 1\end{pmatrix}, \tag{4.49a}$$

$$\eta_-=\eta\begin{pmatrix}\dfrac{\mathrm{i}\sin\gamma}{\cos\gamma+1}\\ 1\end{pmatrix},\qquad \eta_-=\eta\begin{pmatrix}1\\ \dfrac{-\mathrm{i}\sin\gamma}{\cos\gamma+1}\end{pmatrix}, \tag{4.49b}$$

where η is a normalization factor, and the spinors are chosen so that their components will be regular at $\gamma=0$. For electrons, we shall always use the convention that $k=|k|$. For neutrinos, it will prove convenient to let $k=\omega$.

We normalize both the 2-component helicity spinors and Dirac bispinors to unity; the normalization factors are

$$\Lambda=\left(\frac{|\omega|+k}{2|\omega|}\right)^{1/2}\quad\text{and}\quad \eta=\left(\frac{\cos\gamma+1}{2}\right)^{1/2}. \tag{4.50}$$

For the massless neutrino, $m_e=0$, and the Hamiltonian becomes

$$\begin{aligned}H&=-\mathrm{i}\begin{pmatrix}\sigma^k & 0\\ 0 & -\sigma^k\end{pmatrix}\partial_k\\ &=-\mathrm{i}\gamma^5\Sigma^k\partial_k\end{aligned} \tag{4.51}$$

where Σ^k is the kth component of the spin operator (4.36). Using (4.41) we may write (4.51) in terms of the helicity operator

$$H=\gamma^5 hk=q^5 h|\omega| \tag{4.52}$$

where the last equality holds because for massless fields (in natural units) we may choose

$$\omega=k. \tag{4.53}$$

The massless version of the Dirac equation (4.35) or (4.43) agrees with (4.52) and implies

$$\omega\psi=\gamma^5 h|\omega|\psi. \tag{4.54}$$

When the coordinate dependence of (4.42) with $k = \omega$ is assumed, we see from (4.49) that for the 'P-states'

$$\psi_P = \begin{pmatrix} P^a \\ 0 \end{pmatrix},$$

equation (4.54) is satisfied only if P^a is an eigenstate of the helicity matrix

$$\mathcal{m} = \begin{pmatrix} \cos\gamma & -\mathrm{i}\sin\gamma \\ \mathrm{i}\sin\gamma & -\cos\gamma \end{pmatrix}$$

with eigenvalue 1. Similarly, for the 'Q-states',

$$\psi_Q = \begin{pmatrix} 0 \\ Q_{\dot{b}} \end{pmatrix}$$

(4.54) is satisfied only if $Q_{\dot{b}}$ is an eigenstate of $\mathcal{m}$ with eigenvalue -1. Thus we recover the relationship between the sign of ω and the helicity since the above considerations and (4.49) imply

$$P^a = p\left(\frac{\cos\gamma + 1}{2}\right)^{1/2} \begin{pmatrix} 1 \\ \dfrac{\mathrm{i}\sin\gamma}{\cos\gamma + 1} \end{pmatrix} \mathrm{e}^{-\mathrm{i}\chi} \quad \not{c} = \omega/|\omega|$$

$$Q_{\dot{b}} = q\left(\frac{\cos\gamma + 1}{2}\right)^{1/2} \begin{pmatrix} \dfrac{\mathrm{i}\sin\gamma}{\cos\gamma + 1} \\ 1 \end{pmatrix} \mathrm{e}^{-\mathrm{i}\chi} \quad \not{c} = -\omega/|\omega| \qquad (4.55)$$

for massless neutrinos, where $\chi = \omega(t - y\sin\gamma - z\cos\gamma)$. The factors p and q are the amplitudes of the perturbations. Thus the constant spinor amplitudes p^a and $q_{\dot{b}}$ are given by

$$p^a = p\left(\frac{\cos\gamma + 1}{2}\right)^{1/2} \begin{pmatrix} 1 \\ \dfrac{\mathrm{i}\sin\gamma}{\cos\gamma + 1} \end{pmatrix}$$

and

$$q_{\dot{b}} = q\left(\frac{\cos\gamma + 1}{2}\right)^{1/2} \begin{pmatrix} \dfrac{\mathrm{i}\sin\gamma}{\cos\gamma + 1} \\ 1 \end{pmatrix}. \qquad (4.56)$$

4.6 Transformation of plane waves; normal mode expansion

We now transform the solutions (4.55) in flat spacetime in the tetrad given by (4.51) to spherical coordinates (which describe the Kerr metric in Boyer–

Lindquist coordinates as $r \to \infty$) in the asymptotic form of the Kinnersley tetrad. Denoting the flat spacetime solutions with the subscript F and the asymptotic Kerr solutions by the subscript K we have

$$
\begin{aligned}
{}_K P^a &= \rho^a{}_{\underline{a}\mathrm{F}} P^{\underline{a}}, \\
{}_K Q_{\dot{b}} &= \rho^{\underline{\dot{b}}}{}_{\dot{b}\mathrm{F}} Q_{\underline{\dot{b}}},
\end{aligned} \tag{4.57}
$$

where indices of the flat spacetime spinors are underlined. The matrix ρ can be determined by solving the equations

$$
{}_{\mathrm{F}}\sigma^{\underline{\mu}}_{\underline{a}\underline{\dot{b}}} = \frac{\partial X^{\underline{\mu}}}{\partial X^{\nu}}\,{}_{\mathrm{K}}\sigma^{\nu}_{a\dot{b}}\,\rho^a{}_{\underline{a}}\,\bar{\rho}^{\dot{b}}{}_{\underline{\dot{b}}}, \tag{4.58}
$$

where

$$
\begin{aligned}
{}_{\mathrm{K}}\sigma^{\mu}_{0\dot{0}} &\simeq (1, 1, 0, 0), \\
{}_{\mathrm{K}}\sigma^{\mu}_{0\dot{1}} &\simeq (0, 0, a, \mathrm{i}/\sin\theta)/(\sqrt{2}r), \\
{}_{\mathrm{K}}\sigma^{\mu}_{1\dot{1}} &\simeq \tfrac{1}{2}(1, -1, 0, 0),
\end{aligned} \tag{4.59}
$$

in coordinates

$$
X^{\nu} = (t, r, \theta, \phi), \tag{4.60}
$$

subject to the condition that the spin-metric is invariant under the transformation, i.e., that

$$
\varepsilon_{\underline{ab}} = \varepsilon_{ab}\rho^a{}_{\underline{a}}\rho^b{}_{\underline{b}} = \varepsilon_{ab}. \tag{4.61}
$$

The components of ρ are determined up to an overall sign to be

$$
\begin{aligned}
\rho^0{}_{\underline{0}} &= \left(\frac{1+\cos\theta}{2\sqrt{2}}\right)^{1/2} \mathrm{e}^{\mathrm{i}\phi/2} = \rho^{\underline{1}}{}_1 \\
\rho^1{}_{\underline{0}} &= -\left(\frac{1-\cos\theta}{\sqrt{2}}\right)^{1/2} \mathrm{e}^{\mathrm{i}\phi/2} = -\rho^{\underline{1}}{}_0 \\
\rho^0{}_{\underline{1}} &= \left(\frac{1-\cos\theta}{2\sqrt{2}}\right)^{1/2} \mathrm{e}^{-\mathrm{i}\phi/2} = -\rho^{0}{}_1 \\
\rho^1{}_{\underline{1}} &= \left(\frac{1+\cos\theta}{\sqrt{2}}\right)^{1/2} \mathrm{e}^{-\mathrm{i}\phi/2} = \rho^{\underline{0}}{}_0,
\end{aligned} \tag{4.62}
$$

where the symbols on the extreme right in (4.62) are the components of $(\rho)^{-1}$. As a check on (4.62) the NP quantities ϕ_0 and ϕ_2 were transformed (as combinations of the Maxwell spinor components $\mathrm{F}_{abc\dot{d}}$) using the ρ-matrix, and compared to those determined by projecting the tensor $F_{\mu\nu}$ onto the appropriate tetrads; it was found that the above choice of signs is appropriate.

The transformation of (4.57) is written explicitly as

$$
\begin{aligned}
{}_{\mathrm{K}}P^0 &= \left(\frac{1+\cos\theta}{2\sqrt{2}}\right)^{1/2} \mathrm{e}^{\mathrm{i}\phi/2}\,{}_{F}P^0 + \left(\frac{1-\cos\theta}{2\sqrt{2}}\right)^{1/2} \mathrm{e}^{-\mathrm{i}\phi/2}\,{}_{F}P^1,\\
{}_{\mathrm{K}}P^1 &= -\left(\frac{1-\cos\theta}{\sqrt{2}}\right)^{1/2} \mathrm{e}^{\mathrm{i}\phi/2}\,{}_{\mathrm{F}}P^0 + \left(\frac{1+\cos\theta}{\sqrt{2}}\right)^{1/2} \mathrm{e}^{-\mathrm{i}\phi/2}\,{}_{\mathrm{F}}P^1,\\
{}_{\mathrm{K}}Q_{\dot{1}} &= -\left(\frac{1+\cos\theta}{\sqrt{2}}\right)^{1/2} \mathrm{e}^{\mathrm{i}\phi/2}\,{}_{\mathrm{F}}Q_{\dot{0}} - \left(\frac{1-\cos\theta}{\sqrt{2}}\right)^{1/2} \mathrm{e}^{-\mathrm{i}\phi/2}\,{}_{\mathrm{F}}Q_{\dot{1}},\\
{}_{\mathrm{K}}Q_{\dot{0}} &= -\left(\frac{1-\cos\theta}{2\sqrt{2}}\right)^{1/2} \mathrm{e}^{\mathrm{i}\phi/2}\,{}_{\mathrm{F}}Q_{\dot{0}} + \left(\frac{1+\cos\theta}{2\sqrt{2}}\right)^{1/2} \mathrm{e}^{-\mathrm{i}\phi/2}\,{}_{\mathrm{F}}Q_{\dot{1}}, \qquad (4.63)
\end{aligned}
$$

where ${}_{\mathrm{F}}Q_{\dot{b}}$ and ${}_{\mathrm{F}}P^a$ depend on the coordinates as (cf (4.55))

$$
\begin{aligned}
{}_{\mathrm{F}}Q_{\dot{b}} &= q_{\dot{b}}\mathrm{e}^{-\mathrm{i}\chi}\\
{}_{\mathrm{F}}P^a &= p^a \mathrm{e}^{-\mathrm{i}\chi}, \qquad (4.64)
\end{aligned}
$$

with

$$\chi = \omega t - \omega r(\sin\gamma \sin\theta \sin\phi + \cos\gamma \cos\theta) \qquad (4.65)$$

and $q_{\dot{b}}$ and p^a are constant 2-spinors. We refer to the decomposition given by Chandrasekhar ((4.23) above) to expand the wave functions in the appropriate spin-weighted spheroidal harmonics:

$$
\begin{aligned}
{}_{\mathrm{K}}P^0 &\simeq \sum_{lm} {}_{-1/2}C_{lm-1/2}Z_l^m(\theta,\phi;a\omega)\\
{}_{\mathrm{K}}P^1 &\simeq \sum_{lm} {}_{1/2}C_{lm1/2}Z_l^m(\theta,\phi;a\omega)\\
{}_{\mathrm{K}}Q_{\dot{0}} &\simeq \sum_{lm} {}_{-1/2}\hat{C}_{lm1/2}Z_l^m(\theta,\phi;a\omega)\\
{}_{\mathrm{K}}Q_{\dot{1}} &\simeq \sum_{lm} {}_{1/2}\hat{C}_{lm-1/2}Z_l^m(\theta,\phi;a\omega). \qquad (4.66)
\end{aligned}
$$

(We assume for these expansions that $m_e = 0$; the orthonormality of the angular functions for $m_e \neq 0$ is not assured.)

The orthonormality of the spin-weighted spheroidal harmonics allows the evaluation of the expansion coefficients for the neutrino waves, as it did for electromagnetic or linearized gravitational waves. In the latter two cases we could expand a potential (A_μ or $h_{\mu\nu}$), retain terms in the expansion coefficients only to order $(1/r)$, and then use formulae in table 2.1 to determine the relevant NP quantities from the potential. In the case of neutrinos, we must work directly with the NP quantities, and so must retain enough terms to match the asymptotic power-law behavior prescribed by the peeling theorem (see Sachs (1964), or Pirani (1965), for reviews, and NP), or equivalently – for perturbations of the Kerr metric – by the asymptotic behavior of the Teukolsky radial functions.

Chandrasekhar's decomposition indicates that the coefficients in (4.66) must have the following asymptotic behavior:

$$
\begin{aligned}
{}_{-1/2}C_{lm},\ {}_{+1/2}\hat{C}_{lm} &\propto \frac{{}_{-1/2}R_{lm}}{r} \simeq \frac{\mathrm{e}^{-\mathrm{i}\omega r}}{r^2} \quad \text{or} \quad \frac{\mathrm{e}^{\mathrm{i}\omega r}}{r},\\
{}_{1/2}C_{lm},\ {}_{-1/2}\hat{C}_{lm} &\propto {}_{1/2}R_{lm} \simeq \frac{\mathrm{e}^{-\mathrm{i}\omega r}}{r} \quad \text{or} \quad \frac{\mathrm{e}^{\mathrm{i}\omega r}}{r^2}.
\end{aligned}
\tag{4.67}
$$

Eventually we must evaluate terms to one order in r^{-1} beyond the leading order, in these expressions, thus to order $(1/r^2)$ in the expansion coefficients in (4.66). A simpler approach, however, is to evaluate

$$
\begin{aligned}
{}_sC_{lm}(\gamma) = \iint \Bigg[&(-2s)\left(\frac{1-2s\cos\theta}{(3/2-s)\sqrt{2}}\right)^{1/2} p^0 \mathrm{e}^{-\mathrm{i}(\chi-\phi/2)}\\
&+\left(\frac{1+2s\cos\theta}{(3/2-s)\sqrt{2}}\right)^{1/2} p^1 \mathrm{e}^{-\mathrm{i}(\chi+\phi/2)}\Bigg] {}_sZ_l^m(\theta,\phi;a\omega)\,\mathrm{d}\Omega
\end{aligned}
\tag{4.68}
$$

which is obtained from (4.63), (4.64) and (4.66) only to leading order and then use the field equations to extract the subdominant order. This is analogous to the discussion of the gravitational field variables given in section 3.6. Note that ${}_s\hat{C}_{lm}$ is obtained from ${}_sC_{lm}$ by the algebraic replacements

$$
\begin{aligned}
{}_{-1/2}\hat{C}_{lm} &= -\sqrt{2}\,{}_{-1/2}C_{lm}(p^0 \to q_0, p^1 \to q_1),\\
{}_{1/2}\hat{C}_{lm} &= (1/\sqrt{2})\,{}_{1/2}C_{lm}(p^0 \to q_0, p^1 \to q_1).
\end{aligned}
\tag{4.69}
$$

From (A11) we obtain

$$
\begin{aligned}
{}_sC_{lm}(\gamma) = \frac{2\pi}{\mathrm{i}\omega r}\left(\frac{\sqrt{2}}{3-2s}\right)^{1/2} &[\mathrm{i}^{-m+1/2}\mathrm{e}^{-\mathrm{i}\omega(t-r)}F_p(\gamma,s)\,{}_sS_l^m(\gamma;a\omega)\\
&+\mathrm{i}^{m-1/2}\mathrm{e}^{-\mathrm{i}\omega(t+r)}G_p(\pi-\gamma,s)\,{}_sS_l^m(\pi-\gamma;a\omega)],
\end{aligned}
\tag{4.70}
$$

where

$$
\begin{aligned}
F_p(\theta,s) &= -2s(1-2s\cos\theta)^{1/2}p^0 - \mathrm{i}(1+2s\cos\theta)^{1/2}p^1,\\
G_p(\theta,s) &= -2s(1-2s\cos\theta)^{1/2}p^0 + \mathrm{i}(1+2s\cos\theta)^{1/2}p^1.
\end{aligned}
\tag{4.71}
$$

Using the relation between p^0 and p^1 given by (4.56) we obtain

$$
F_p(\gamma,\tfrac{1}{2}) = G_p(\pi-\gamma,-\tfrac{1}{2}) = 0. \tag{4.72}
$$

To determine the subdominant parts of ${}_sC_{lm}$, we exploit (4.22) with

$\mu_e = m_e/\sqrt{2} = 0$, and (4.66). Specifically, we let

$$P_0 = \left(\frac{2^{1/4}\pi}{i\omega r}\right)\sum_{lm}\left[(-i)^{m-1/2}e^{-i\omega(t-r^*)}F_p(\gamma, -\tfrac{1}{2})_{-1/2}S_l^m(\gamma; a\omega) + \frac{C^0}{r}e^{-i\omega(t+r^*)}\right]{}_{1/2}Z_l^m(\theta, \phi; a\omega) \tag{4.73}$$

and

$$P^1 = \left(\frac{2^{3/4}\pi}{i\omega r}\right)\sum_{lm}\left[\frac{C^1}{r}e^{-i\omega(t-r^*)} - (i)^{m-1/2}e^{-i\omega(t+r^*)} \times G_p(\pi-\gamma, \tfrac{1}{2})_{1/2}S_l^m(\pi-\gamma; a\omega)\right]{}_{1/2}Z_l^m(\theta, \phi; a\omega) \tag{4.74}$$

where C^0 and C^1 are as yet undetermined functions, and we have used (4.72). Because of the linearity of (4.22), they must be satisfied separately by the ingoing parts of P^0, P^1, and separately by the outgoing parts of P^0, P^1. This simplifies the analysis which is even further simplified by considering equations only in the very large r limit. Insert the ingoing piece of P^0, P^1 into the first of (4.22), and the outgoing piece of P^0, P^1 into the second of (4.22) obtaining, as $r \to \infty$

$$(\partial_r - i\omega)P^0 = -\frac{1}{r\sqrt{2}}{}_{1/2}\mathscr{L}P^1, \tag{4.75}$$

$$(\partial_r + i\omega)P^1 = \frac{\sqrt{2}}{r}{}_{1/2}\mathscr{L}^\dagger P^0, \tag{4.76}$$

where the behavior is 'ingoing' in (4.75) and 'outgoing' in (4.76).

The angular operators on the right are evaluated by the $m_e \to 0$ limit of (4.24), yielding

$$C^0 = \frac{\lambda^{1/2}}{-2i\omega}(i)^{m-1/2}G_p(\pi-\gamma, \tfrac{1}{2})_{1/2}S_l^m(\pi-\gamma; a\omega),$$

$$C^1 = \frac{\lambda^{1/2}}{2i\omega}(-i)^{m-1/2}F_p(\gamma, -\tfrac{1}{2})_{-1/2}S_l^m(\gamma; a\omega). \tag{4.77}$$

Thus comparison of (4.74) and (4.77) with (4.66) indicates that the coefficient ${}_sC_{lm}$ is given by

$${}_sC_l^m(\gamma) = \left(\frac{2\pi}{i\omega r}\right)\left(\frac{\sqrt{2}}{3-2s}\right)^{1/2}\left\{(-i)^{m-1/2}e^{-i\omega(t-r^*)} \times \left[F_p(\gamma, -\tfrac{1}{2})_{-1/2}S_l^m(\gamma; a\omega)\left(\delta_{s,-1/2} + \frac{\delta_{s,1/2}\lambda^{1/2}}{2i\omega r}\right)\right]\right.$$

$$+(\mathrm{i})^{m-1/2}\mathrm{e}^{-\mathrm{i}\omega(t+r^*)}\Bigg[-G_p(\pi-\gamma,\tfrac{1}{2})S_l^m(\pi-\gamma;a\omega)$$

$$\times\left(\delta_{s,1/2}+\frac{\delta_{s,-1/2}\lambda^{1/2}}{2\mathrm{i}\omega r}\right)\Bigg]\Bigg\}. \tag{4.78}$$

Axial incidence is obtained by taking $\gamma\to 0$ in (4.78).

Equations (4.66), (4.69), (4.71), and (4.78) taken together complete the expansion of neutrino plane waves in the background of the Kerr metric. The expansions so determined must now be matched to the 'normal mode' expansions of the neutrino field determined by (4.23) namely,

$$P^{0\,\mathrm{up}}_{\ \mathrm{dn}}=-\rho\int\mathrm{d}\tilde{\omega}\sum_{lm}K^{\mathrm{up}}_{\mathrm{dn}\,lm\tilde{\omega}}\,{}_{-1/2}R^{\mathrm{up}}_{\mathrm{dn}\,lm\tilde{\omega}}\,{}_{-1/2}Z_l^m(\theta,\phi;a\tilde{\omega})\mathrm{e}^{-\mathrm{i}\tilde{\omega}t},$$

$$P^{1\,\mathrm{up}}_{\ \mathrm{dn}}=\sqrt{2}\int\mathrm{d}\tilde{\omega}\sum_{lm}K^{\mathrm{up}}_{\mathrm{dn}\,lm\tilde{\omega}}\,{}_{1/2}R^{\mathrm{up}}_{\mathrm{dn}\,lm\tilde{\omega}}\,{}_{1/2}Z_l^m(\theta,\phi;a\tilde{\omega})\mathrm{e}^{-\mathrm{i}\tilde{\omega}t},$$

$$Q^{\dot{0}\,\mathrm{up}}_{\ \mathrm{dn}}=-\rho\int\mathrm{d}\tilde{\omega}\sum_{lm}\hat{K}^{\mathrm{up}}_{\mathrm{dn}\,lm\tilde{\omega}}\,{}_{-1/2}R^{\mathrm{up}}_{\mathrm{dn}\,lm\tilde{\omega}}\,{}_{1/2}Z_l^m(\theta,\phi;a\tilde{\omega})\mathrm{e}^{-\mathrm{i}\tilde{\omega}t},$$

$$Q^{\dot{1}\,\mathrm{up}}_{\ \mathrm{dn}}=\sqrt{2}\int\mathrm{d}\tilde{\omega}\sum_{lm}\hat{K}^{\mathrm{up}}_{\mathrm{dn}\,lm\tilde{\omega}}\,{}_{-1/2}R^{\mathrm{up}}_{\mathrm{dn}\,lm\tilde{\omega}}\,{}_{1/2}Z_l^m(\theta,\phi;a\tilde{\omega})\mathrm{e}^{-\mathrm{i}\tilde{\omega}t}. \tag{4.79}$$

These expansions are matched to the asymptotic expansions determined above, and are assumed to be valid at all radii.

We choose the normalizations

$${}_{-1/2}R^{\mathrm{up}}_{lm\tilde{\omega}}\underset{r\to\infty}{=}2^{-3/4}\mathrm{e}^{\mathrm{i}\omega r^*},$$

$${}_{1/2}R^{\mathrm{up}}_{lm\tilde{\omega}}\underset{r\to\infty}{=}\frac{\lambda^{1/2}2^{-3/4}\mathrm{e}^{\mathrm{i}\tilde{\omega}r^*}}{2\mathrm{i}\tilde{\omega}r^2}, \tag{4.80}$$

and

$${}_{-1/2}R^{\mathrm{dn}}_{lm\tilde{\omega}}\underset{r\to\infty}{=}\frac{\lambda^{1/2}2^{-3/4}\mathrm{e}^{-\mathrm{i}\tilde{\omega}r^*}}{2\mathrm{i}\tilde{\omega}r},$$

$$R^{\mathrm{dn}}_{lm\tilde{\omega}}\underset{r\to\infty}{=}-2^{-3/4}\frac{\mathrm{e}^{-\mathrm{i}\tilde{\omega}r^*}}{r}.$$

Such normalizations are consistent with the asymptotic forms of (4.24) and (4.25) with $m_e=0$. With the normalizations in (4.80) we obtain

$$K^{\mathrm{up}}_{lm\tilde{\omega}}=\frac{2\pi}{\mathrm{i}\omega}(-\mathrm{i})^{m-1/2}F_p(\gamma,-\tfrac{1}{2})\,{}_{-1/2}S_l^m(\gamma;a\omega)\delta(\omega-\tilde{\omega}),$$

$$K^{\mathrm{dn}}_{lm\tilde{\omega}}=\frac{2\pi}{\mathrm{i}\omega}(\mathrm{i})^{m-1/2}G_p(\pi-\gamma,\tfrac{1}{2})\,{}_{1/2}S_l^m(\pi-\gamma;a\omega)\delta(\omega-\tilde{\omega}), \tag{4.81}$$

$$\hat{K}^{\mathrm{up}}_{lm\tilde{\omega}} = \frac{2\pi}{\mathrm{i}\omega}(-\mathrm{i})^{m-1/2}F_q(\gamma, \tfrac{1}{2})_{1/2}S_l^m(\gamma; a\omega)\delta(\omega - \tilde{\omega}),$$

$$\hat{K}^{\mathrm{dn}}_{lm\tilde{\omega}} = \frac{2\pi}{\mathrm{i}\omega}(\mathrm{i})^{m-1/2}G_q(\pi - \gamma, -\tfrac{1}{2})_{-1/2}S_l^m(\pi - \gamma; a\omega)\delta(\omega - \tilde{\omega}). \quad (4.82)$$

The formulae in (4.82) are obtained by matching (4.79) to (4.66) with the coefficients ${}_sC_l^m(\gamma)$ given by (4.78). For the lower components of the Dirac bispinor, Q^b, we have anticipated results which we now derive. We use (4.69) and (4.78) to obtain ${}_s\hat{C}_{lm}$ and match the last pair of (4.82). First we rewrite ${}_sC_{lm}$ in the form

$$\begin{aligned}{}_sC_{lm}(\gamma) = \left(\frac{2\pi}{\mathrm{i}\omega r}\right)\left(\frac{\sqrt{2}}{3-2s}\right)^{1/2}\Bigg[&(-\mathrm{i})^{m-1/2}\mathrm{e}^{-\mathrm{i}\omega(t-r^*)}\Bigg(F_p(\gamma, s)_sS_l^m(\gamma; a\omega)\\ &+\frac{\lambda\!\!\!^{-1/2}}{2\mathrm{i}\omega r}F_p(\gamma, -s)_{-s}S_l^m(\gamma; a\omega)\Bigg) - (\mathrm{i})^{m-1/2}\mathrm{e}^{-\mathrm{i}\omega(t+r^*)}\\ &\times\Bigg(G_p(\pi-\gamma, s)_sS_l^m(\pi-\gamma; a\omega)\\ &-\frac{\lambda\!\!\!^{-1/2}}{2\mathrm{i}\omega r}G_p(\pi-\gamma, -s)_{-s}S_l^m(\pi-\gamma; a\omega)\Bigg)\Bigg]\end{aligned} \quad (4.83)$$

and then make the replacements indicated by (4.69). Noting that

$$F_q(\gamma, -\tfrac{1}{2}) = G_q(\pi - \gamma, \tfrac{1}{2}) = 0$$

we obtain

$$\begin{aligned}{}_s\hat{C}_{lm}(\gamma) = (-2s)\left(\frac{2\pi}{\mathrm{i}\omega r}\right)\left(\frac{\sqrt{2}}{3+2s}\right)^{1/2}\Bigg[&(-\mathrm{i})^{m-1/2}\mathrm{e}^{-\mathrm{i}\omega(t-r^*)}\Bigg(F_q(\gamma, s)_sS_l^m(\gamma, a\omega)\\ &+\frac{\lambda\!\!\!^{-1/2}}{2\mathrm{i}\omega r}F_q(\gamma, -s)_{-s}S_l^m(\gamma; a\omega)\Bigg) - (\mathrm{i})^{m-1/2}\mathrm{e}^{-\mathrm{i}\omega(t+r^*)}\\ &\times\Bigg(G_q(\pi-\gamma; s)_sS_l^m(\pi-\gamma; a\omega)\\ &+\frac{\lambda\!\!\!^{-}}{2\mathrm{i}\omega r}G_q(\pi-\gamma, -s)_{-s}S_l^m(\pi-\gamma; a\omega)\Bigg)\Bigg].\end{aligned} \quad (4.84)$$

Inserting (4.84) in the expansions (4.66) for the components Q^b, and comparing the result to (4.79) then yields the last pair of (4.82).

Inserting the formulae for p^a and q_b (4.56) in the definitions (4.71) of F and

G, we obtain

$$\frac{F_p(\gamma, -\frac{1}{2})}{p} = \mathrm{i}\frac{F_q(\gamma, \frac{1}{2})}{q} = \sqrt{2},$$

$$\frac{G_p(\pi - \gamma, \frac{1}{2})}{p} = \mathrm{i}\frac{G_q(\pi - \gamma, -\frac{1}{2})}{q} = -\sqrt{2}. \tag{4.85}$$

Thus, the final form of (4.82) is

$$K^{\mathrm{up}}_{lm\tilde{\omega}} = \frac{2\sqrt{2}\pi p}{\mathrm{i}\omega}(-\mathrm{i})^{m-1/2}{}_{-1/2}S^m_l(\gamma; a\omega)\delta(\omega - \tilde{\omega}),$$

$$K^{\mathrm{dn}}_{lm\tilde{\omega}} = -\frac{2\sqrt{2}\pi p}{\mathrm{i}\omega}(\mathrm{i})^{m-1/2}{}_{1/2}S^m_l(\pi - \gamma; a\omega)\delta(\omega - \tilde{\omega}),$$

$$\hat{K}^{\mathrm{up}}_{lm\tilde{\omega}} = -\frac{2\sqrt{2}\pi q}{\omega}(-\mathrm{i})^{m-1/2}{}_{1/2}S^m_l(\gamma; a\omega)\delta(\omega - \tilde{\omega}),$$

$$\hat{K}^{\mathrm{dn}}_{lm\tilde{\omega}} = \frac{2\sqrt{2}\pi q}{\omega}(\mathrm{i})^{m-1/2}{}_{-1/2}S^m_l(\pi - \gamma; a\omega)\delta(\omega - \tilde{\omega}). \tag{4.86}$$

The neutrino mode amplitudes given in (4.86) differ in form from those corresponding to electromagnetic and linearized gravitational perturbations in that there is no decomposition into amplitudes corresponding to modes of differing parities. Instead, the decomposition is in terms of 'p' modes and 'q' modes. The 'p' and 'q' states may be interpreted in the usual ways (cf. Rose, 1961; Schweber, 1967) as charge conjugate states satisfying the two-component Weyl equation, and as is well known, the two-component equation is not invariant under parity transformations. Whether p^a is called a neutrino and $q^{\dot{a}}$ an anti-neutrino, or vice versa, is a matter of convention.

5
Scattering

5.1 Scattering of scalar waves

Consider first the scalar case. To compute a cross section we define

$$\tilde{K}_{lm\tilde{\omega}} = K^{\text{up}}_{lm\tilde{\omega}} - K^{\text{up}}_{lm\tilde{\omega}} \quad \text{(plane)} \tag{5.1}$$

where $\tilde{K}_{lm\tilde{\omega}}$ is the scattered mode amplitude, and

$$\tilde{\phi} = \int \mathrm{d}\tilde{\omega} \sum_{lm} \tilde{K}_{lm\tilde{\omega}0} R^{\text{up}}_{lm\tilde{\omega}0} Z^m_l(\theta, \phi; a\tilde{\omega}) \mathrm{e}^{-\mathrm{i}\tilde{\omega}t} \tag{5.2}$$

is the scattered wave. The total solution has been normalized so that $K^{\text{dn}}_{lm\tilde{\omega}} = K^{\text{dn}}_{lm\tilde{\omega}}$(plane). The number density is given by the time component of the conserved current

$$j^0 = -\mathrm{i}[(\nabla^0\bar{\phi})\phi - (\nabla^0\phi)\bar{\phi}],$$

so that

$$j^0(\text{plane}) = \omega b^2,$$

$$\tilde{j}^0(\text{scatt}) \underset{r\to\infty}{\simeq} \frac{\omega}{r^2}\left|\sum_{lm}\int \tilde{K}_{lm\tilde{\omega}0} S^m_l(\gamma; a\tilde{\omega})\,\mathrm{d}\tilde{\omega}\right|^2 \tag{5.3}$$

(where b is the amplitude of the scalar wave, cf. (3.20)) and the scattering cross section is

$$\frac{\mathrm{d}\sigma}{\mathrm{d}\Omega} = r^2 \frac{\tilde{j}^0}{j^0(\text{plane})} = \frac{1}{b^2}\left|\sum_{lm} \tilde{K}_{lm\omega0} S^m_l(\gamma; a\omega)\right|^2 \tag{5.4}$$

where the integration over $\tilde{\omega}$ has been performed. The cross section given above is also obtained when one uses the energy and flux densities implied by (cf. Sakurai, 1967; Schweber, 1961)

$$T^{\alpha\beta} = \tfrac{1}{2}(\phi^{;\alpha}\bar{\phi}^{;\beta} - g^{\alpha\beta}\phi^{;\mu}\bar{\phi}_{;\mu}) + \text{complex conjugate.} \tag{5.5}$$

With this result as a guide we now consider scattering for the other integer-spin cases, and then for neutrino scattering.

Notice that the essential step in this process has only been indicated, and must be explicitly carried out. It is the derivation of the total solution with the appropriate boundary conditions. We will present in chapter 7 the results of some numerical integrations of the differential equation. Other techniques are possible. For instance, for the scalar Schwarzschild case, Sanchez (1978a,b) obtains the solution by a technique of repeated analytic continuation upon a solution asymptotically defined at $r = \infty$.

5.2 Scattering of electromagnetic waves

For the electromagnetic problem, the asymptotic fields (either the A_m, $A_{\bar{m}}$; or the fields ϕ_0, ϕ_2) are compared to the ones arising with scattering boundary conditions when the 'down' parts are normalized to correspond to the 'down' part of the plane wave. As for the scalar case, the difference in the 'up' part constitutes the scattered wave.

Expressing the scattered wave as a mode sum leads to

$$A_{\bar{m}}^{\text{scatt}} = \int_{-\infty}^{\infty} \mathrm{d}\tilde{\omega} \sum_{lmP} \tilde{K}_{lm\tilde{\omega}P} R_{lm\tilde{\omega}-1}^{\text{up}} Z_l^m(\theta, \phi; a\tilde{\omega}) \mathrm{e}^{-\mathrm{i}\tilde{\omega}t}, \tag{5.6}$$

where

$$\tilde{K}_{lm\tilde{\omega}P} = K_{lm\tilde{\omega}P}^{\text{up}} - K_{lm\tilde{\omega}P}^{\text{up}}(\text{plane})$$

is the scattered mode amplitude. We use the fact that $\tilde{K}_{lm\tilde{\omega}P}$ must satisfy the reality relation (3.45), to express it as

$$\tilde{K}_{lm\tilde{\omega}P} = \tilde{k}_{lm\tilde{\omega}P}\delta(\omega - \tilde{\omega}) + \tilde{k}_{l-m-\tilde{\omega}P}\delta(\omega + \tilde{\omega}),$$

where

$$P\bar{\tilde{k}}_{lm\tilde{\omega}P} = \tilde{k}_{l-m-\tilde{\omega}P}(-1)^{l+m+1}. \tag{5.8}$$

(since $|s| = 1$ for electromagnetism.) From the asymptotic form of $\bar{m}$ (cf (2.21)), we have for the scattered wave

$$\tilde{A}_{\bar{m}} \underset{r\to\infty}{\sim} \frac{1}{\sqrt{2}}(\tilde{A}_{\hat{\theta}} - \mathrm{i}\tilde{A}_{\hat{\phi}}), \tag{5.9}$$

a purely transverse radially propagating wave. The notation $\hat{\theta}$, $\hat{\phi}$ indicates component along a unit vector in the θ or ϕ direction. We thus may calculate the asymptotic electric field:

$$\tilde{E}_{\hat{\theta}} - \mathrm{i}\tilde{E}_{\hat{\phi}} = \sqrt{2}(\partial\tilde{A}_{\bar{m}}/\partial t). \tag{5.10}$$

Because these are transverse radiation fields, their energy density $|T_{00}|$ is equal to the magnitude of the Poynting vector, $|\mathbf{S}| = T_{0i}n^i$, where $i = 1, 2, 3$, n^i is a unit vector directed radially outward. Hence the asymptotic energy density in the outgoing scattered wave is

$$|T_{00}|_{\text{scatt}} = \frac{1}{4\pi}\left\langle \left|\frac{\partial \tilde{A}_\theta}{\partial t}\right|^2 + \left|\frac{\partial \tilde{A}_\phi}{\partial t}\right|^2 \right\rangle$$

$$= \frac{1}{4\pi}\left\langle \left|\frac{\partial A_{\bar{m}}}{\partial t}\right|^2 \right\rangle, \tag{5.11}$$

where the brackets $\langle\,\rangle$ denote time average (over one cycle).

The energy flux in the incident (inc) wave is: (cf (3.28), (3.29))

$$|T_{00}|_{\text{inc}} = \langle A^2[(\omega\cos\omega\chi)^2 + (\omega\sin\omega\chi)^2]\rangle = \frac{\omega^2 A^2}{8\pi}. \tag{5.12}$$

The cross section is

$$\frac{\mathrm{d}\sigma}{\mathrm{d}\Omega} = \frac{r^2|T_{00}|_{\text{scatt}}}{|T_{00}|_{\text{inc}}} \tag{5.13}$$

$$= 2r^2|A|^{-2}\left(\left|\sum_{lmP} \tilde{k}_{lm\omega P-1}S_l^m(\theta; a\tilde{\omega})\right|^2\right.$$

$$\left. + \left|\sum_{lmP}(-1)^{l+m+s}P\tilde{k}_{lm\omega P-1}S_l^m(\pi-\theta; a\tilde{\omega})\right|^2\right).$$

Teukolsky (1973) gives an alternative discussion in which $\mathrm{d}\sigma/\mathrm{d}\Omega$ is defined in terms of the energy flux and can be expressed in terms of $|\tilde{\phi}_2|^2$ alone.

5.3 Scattering of gravitational radiation

Now consider the gravitational-wave case. The discussion below is adapted from Matzner & Ryan (1978).

The scattering cross section is related to the constants

$$\tilde{K}_{lm\tilde{\omega}P} \equiv K_{lm\tilde{\omega}P}{}^{\text{up}}(\text{scatt}) = K_{lm\tilde{\omega}P}{}^{\text{up}} - K_{lm\tilde{\omega}P}{}^{\text{up}}(\text{plane}) \tag{5.14}$$

through the energy flux formula

$$\frac{\mathrm{d}E}{\mathrm{d}t\,\mathrm{d}\Omega} \equiv \frac{\mathrm{d}E^{\text{up}}}{\mathrm{d}t\,\mathrm{d}\Omega}(\text{scatt}) = \lim_{r\to\infty} r^2 T^r{}_t.$$

However, for the gravitational wave case, no stress tensor can be defined. Instead, we take as $T^\mu{}_\nu$ the Isaacson (1968) effective stress tensor formed from $\tilde{h}_{\mu\nu} \equiv h_{\mu\nu}{}^{\text{up}}$ (scatt) in a transverse traceless gauge. Specifically, one has

$$\frac{\mathrm{d}\tilde{E}}{\mathrm{d}t\,\mathrm{d}\Omega} = -\lim_{r\to\infty}\frac{r^2}{32\pi}\left\langle \frac{\partial}{\partial t}\tilde{h}_{mm}\frac{\partial}{\partial r}\tilde{h}_{\bar{m}\bar{m}} + \text{complex conjugate}\right\rangle, \tag{5.15}$$

with the angular brackets denoting an average over several wavelengths. In a transverse traceless gauge, $\tilde{h}_{\bar{m}\bar{m}}$ has the asymptotic form (cf. section 3.6 and Chrzanowski (1975))

$$\tilde{h}_{\bar{m}\bar{m}} = \int_{-\infty}^{\infty} \sum_{lmP} \mathrm{d}\tilde{\omega} \tilde{K}_{lm\tilde{\omega}P} \frac{\mathrm{e}^{\mathrm{i}\tilde{\omega}(r^* - t)}}{r} {}_{-2}Z_l^m(\theta, \phi; a\tilde{\omega}). \tag{5.16}$$

Carrying out the integration over the frequency and using the symbol $\tilde{k}_{lm\tilde{\omega}P}$ defined by

$$\tilde{K}_{lm\tilde{\omega}P} = \tilde{k}_{lm\tilde{\omega}P}\delta(\omega - \tilde{\omega}) + \tilde{k}_{l-m-\tilde{\omega}P}\delta(\omega + \tilde{\omega}) \tag{5.17}$$

together with the symmetries of the angular function, we obtain

$$\tilde{h}_{\bar{m}\bar{m}} \to \sum_{lmP} \left(\tilde{k}_{lm\tilde{\omega}P} \frac{\mathrm{e}^{\mathrm{i}\omega(r^* - t)}}{r} {}_{-2}S_l^m(\theta; a\omega)\mathrm{e}^{\mathrm{i}m\phi} \right.$$
$$\left. + (-1)^{l+m+s} P \tilde{k}^*_{lm\tilde{\omega}P} \frac{\mathrm{e}^{-\mathrm{i}\omega(r^* - t)}}{r} {}_{-2}S_l^m(\pi - \theta; a\omega)\mathrm{e}^{-\mathrm{i}m\phi} \right). \tag{5.18}$$

To obtain (5.18) we used the metric reality condition

$$K_{lm\tilde{\omega}P} = P\bar{K}_{l-m-\tilde{\omega}P}(-1)^{l+m+s} \tag{5.19}$$

which implies the same relation for $\tilde{k}_{lm\tilde{\omega}P}$. It follows that

$$\frac{\mathrm{d}\tilde{E}}{\mathrm{d}t\,\mathrm{d}\Omega} = \lim_{r\to\infty} \frac{\omega^2 r^2}{16\pi} \langle |\tilde{h}_{\bar{m}\bar{m}}|^2 \rangle = \frac{\omega^2}{16\pi} \left(\left| \sum_{lmP} \tilde{k}_{lm\omega P} \, {}_{-2}S_l^m(\theta, a\omega) \right|^2 \right.$$
$$\left. + \left| \sum_{lmP} (-1)^{l+m+s} P \tilde{k}_{lm\omega P} \, {}_{-2}S_l^m(\pi - \theta, a\omega) \right|^2 \right), \tag{5.20}$$

where terms with time dependent factors have been averaged out.

To find the scattering cross section, we divide $\mathrm{d}\tilde{E}/\mathrm{d}t\,\mathrm{d}\Omega$ by

$$\frac{\mathrm{d}E}{\mathrm{d}t\,\mathrm{d}A}(\text{plane}) = \lim_{z\to-\infty} \cos\gamma T^z{}_t + \lim_{y\to-\infty} \sin\gamma T^y{}_t.$$

However, since the energy per unit area in the incident wave is independent of the angle of incidence, we use the expression from Matzner & Ryan (1978) for the axially incident case

$$\frac{\mathrm{d}E}{\mathrm{d}t\,\mathrm{d}A}(\text{plane}) = \lim_{z\to-\infty} T^z{}_t$$
$$= -\lim_{z\to-\infty} \frac{1}{16\pi} \left\langle \frac{\partial}{\partial t}(h_{xx})\frac{\partial}{\partial z}(h_{xx}) + \frac{\partial}{\partial t}(h_{xy})\frac{\partial}{\partial z}(h_{xy}) \right\rangle$$
$$= \frac{h^2\omega^2}{16\pi}, \tag{5.21}$$

where we use equation (3.58) for h_{xx} and h_{xy}. Hence, from (5.20) and (5.21), we find that the scattering cross section is given by

$$\frac{d\sigma}{d\Omega} = \left(\frac{dE}{dt\, d\Omega}\right) \Big/ \left(\frac{dE}{dt\, dA}(\text{plane})\right) \tag{5.22}$$

$$= \frac{1}{h^2}\left(\left|\sum_{lmP} \tilde{k}_{lm\omega P}\, {}_{-2}S_l^m(\theta, a\omega)\right|^2 + \left|\sum_{lmP} (-1)^{l+m+s} P \tilde{k}_{lm\omega P}\, {}_{-2}S_l^m(\pi - \theta, a\omega)\right|^2\right).$$

5.4 Scattering of Neutrinos

Now we derive a formula for the neutrino scattering cross section in terms of the scattered mode amplitudes. The stress-energy spinor for electrons is (Güven, 1977)

$$T^{a\dot{a}b\dot{b}} = -\tfrac{1}{2}\mathrm{i}(P^a \nabla^{b\dot{b}} P^{\dot{a}} - P^{\dot{a}} \nabla^{b\dot{b}} P^a + P^b \nabla^{a\dot{a}} P^{\dot{b}} - P^{\dot{b}} \nabla^{a\dot{a}} P^b + Q^{\dot{a}} \nabla^{b\dot{b}} Q^a - Q^a \nabla^{b\dot{b}} Q^{\dot{a}} + Q^{\dot{b}} \nabla^{a\dot{a}} Q^b - Q^b \nabla^{a\dot{a}} Q^{\dot{b}}). \tag{5.23}$$

By taking projections

$$T^{\mu\nu} = \sigma^{\mu}_{a\dot{a}} \sigma^{\nu}_{b\dot{b}} T^{a\dot{a}b\dot{b}} \tag{5.24}$$

in one or the other of the dyad (tetrad) bases given by (4.31) or (4.59) one finds that

$$T^{00}(\text{plane wave}) = (2\omega/\sqrt{2})(q^2 + p^2) \tag{5.25}$$

and, to leading order in $(1/r)$,

$$\tilde{T}^{00}(\text{scatt}) = 2\omega(\tilde{Q}^0 \tilde{Q}^{\dot{0}} + \tilde{P}^0 \tilde{P}^{\dot{0}}) \tag{5.26}$$

where the tilde ($\sim$) denotes the scattered wave (expressed relative to the Kinnersley tetrad (4.59)). Similarly, we find for the net number current

$$J^{\mu} = \sigma^{\mu}_{a\dot{a}}(P^a P^{\dot{a}} + Q^a Q^{\dot{a}}) \tag{5.27}$$

$$J_{\text{in}}(\text{plane}) = \cos\gamma\, J_z(\text{plane}) + \sin\gamma\, J_y(\text{plane})$$

$$= \frac{1}{\sqrt{2}}(p^2 + q^2) \tag{5.28}$$

and

$$\tilde{J}_r(\text{scatt}) \underset{r\to\infty}{=} (\tilde{P}^0 \tilde{P}^{\dot{0}} + \tilde{Q}^0 \tilde{Q}^{\dot{0}}). \tag{5.29}$$

The scattering cross sections are

$$\frac{d\sigma}{d\Omega_E} = r^2 \frac{\langle \tilde{T}^{00} \rangle}{\langle T^{00}(\text{plane}) \rangle} = \left(\frac{\sqrt{2}}{q^2 + p^2} \right) \langle \tilde{T}^{00} \rangle \tag{5.30}$$

and

$$\frac{d\sigma}{d\Omega_N} = r^2 \frac{\langle \tilde{J}_r \rangle}{\langle J_{\text{inc}}(\text{plane}) \rangle} = \left(\frac{\sqrt{2}}{p^2 + q^2} \right) \langle \tilde{J}_r \rangle \tag{5.31}$$

where angle brackets denote time averaging. Expanding $\tilde{P}^0$ and $\tilde{Q}^{\dot{0}}$ in mode sums (in the limit of large r)

$$\tilde{P}^0 \sim \frac{1}{r} \int d\tilde{\omega} \sum_{lm} \tilde{k}_{lm\tilde{\omega} - 1/2} Z_l^m(\theta, \phi; a\tilde{\omega}) e^{-i\tilde{\omega}(t - r^*)}$$

$$\tilde{Q}^{\dot{0}} \sim -\frac{1}{r} \int d\tilde{\omega} \sum_{lm} \hat{k}_{lm\tilde{\omega} 1/2} Z_l^m(\theta, \phi; a\tilde{\omega}) e^{-i\tilde{\omega}(t - r^*)} \tag{5.32}$$

we find that the two cross sections are identical and are given by

$$\frac{d\sigma}{d\Omega} = \left(\frac{\sqrt{2}}{q^2 + p^2} \right) \left(\left| \sum_{lm} \tilde{\hat{k}}_{lm\omega 1/2} S_l^m(\theta; a\omega) \right|^2 + \left| \sum_{lm} \tilde{k}_{lm\omega - 1/2} S_l^m(\theta; a\omega) \right|^2 \right). \tag{5.33}$$

The scattering mode amplitudes are given by

$$\tilde{\hat{K}}_{lm\omega} = \hat{K}^{\text{up}}_{lm\omega} - \hat{K}^{\text{up}}_{lm\omega}(\text{plane}) \tag{5.34}$$

and

$$\tilde{K}_{lm\omega} = K^{\text{up}}_{lm\omega} - K^{\text{up}}_{lm\omega}(\text{plane}) \tag{5.35}$$

where the amplitudes $\hat{K}^{\text{up}}_{lm\omega}$ and $K^{\text{up}}_{lm\omega}$ are determined by solving the radial equation.

5.5 Absorption cross section

For all the situations of interest the total absorption cross section is computed in a similar manner. We compute the outgoing flux using $\tilde{k}_{lm\tilde{\omega}P}$, or $\hat{k}_{lm\omega}$ for neutrinos, then subtract the flux calculated using $k_{lm\tilde{\omega}P}{}^{\text{up}}$ (plane wave) (or $\tilde{\hat{k}}_{lm\omega}$). The result when integrated over solid angle and divided by the incident flux gives the inelastic cross section:

$$\sigma_{\text{inel}} = \Sigma_{lm\tilde{\omega}P} \sigma_{\text{inel}}(lm\tilde{\omega}P), \tag{5.36}$$

$$\sigma_{\text{inel}}(lm\tilde{\omega}P) = (|k_{lm\tilde{\omega}P}{}^{\text{up}}(\text{plane wave})|^2 - |\tilde{k}_{lm\tilde{\omega}P}|^2)/h^2, \tag{5.37}$$

or the equivalent neutrino form, where h stands generically for the amplitude b of the scalar waves, the amplitude of $2^{-1/2}(p^2+q^2)$ of neutrinos, the amplitude $|A|$ of the vector potential, and the amplitude h of the metric perturbation. Positive σ_{inel} corresponds to absorption; negative σ_{inel} corresponds to superradiance. (Chandrasekhar & Detweiler (1977) have shown that superradiance does not occur in neutrino scattering.)

5.6 Scattering and phase shifts

It is possible in all cases to write the perturbation problem in terms of quantities that have similar $(1/r)$ falloff for their up and down modes at ∞; cf Chandrasekhar & Detweiler (1977). Then, the total wave's up part can be written

$$\hat{k}_{lm\omega P} = k^{\text{up}}_{lm\omega P}(\text{plane wave})(e^{2i\delta_l} - 1)$$

where δ_l is real if there is no absorption; complex otherwise. (For neutrinos we consider a decomposition in helicity rather than parity.)

Then

$$\frac{d\sigma}{d\Omega} = r^2 \frac{\langle T^{00} \rangle}{\langle T^{00}_{\text{plane}} \rangle} = \frac{1}{h^2} \left| \sum_{lmP} k^{\text{up}}_{lm\omega P}(\text{plane})(e^{2i\delta_{lP}} - 1)\,{}_{-s}S^m_l(\theta; a\omega) \right|^2$$
$$+ \frac{1}{h^2} \left| \sum_{lmP} (-1)^{l+m+s} P k^{\text{up}}_{lm\omega P}(\text{plane})(e^{2i\delta_{lP}} - 1)\,{}_{-s}S^m_l(\pi - \theta; a\omega)] \right|^2 \tag{5.38}$$

where h, as above, is representative of the amplitude of the wave in the different cases considered.

6
Limiting cross sections

In the previous chapters we have alluded to the analogies between massless wave black hole scattering, and quantum mechanical scattering, especially by the Coulomb potential. In this chapter we will systematize these approximations. The low and high frequency limiting cross sections considered here will aid the interpretation of the numerical results of the next few chapters.

6.1 Low frequency cross sections

Since we consider scattering processes, we will concentrate on low frequency behavior related to the field at infinity. That means, for instance, that any possible low frequency structure near the horizon will be ignored, so long as it has no effect at infinity.

From this viewpoint of scattering, one expects only a simple angular pattern at low frequencies; all the detail should wash out. As we show immediately below, the low frequency limit is actually a limit in $(M\omega/l) \to 0$ for large l (l is total angular momentum). Hence small l, in particular $l = 0$ or $l = \frac{1}{2}$, leads to anomalous cases.

6.1.1 Low frequency scalar cross sections

We begin by discussing the scalar case, which is much simpler than the others. Into the scalar ($s = 0$) radial equation (2.27), introduce the change of dependent variable

$$R = \Delta^{-1/2} u,$$

to obtain

$$\frac{d^2u}{dr^2} + \left(\frac{K^2}{\Delta} - (E + a^2\omega^2 - 2am\omega) - \frac{(a^2 - M^2)}{\Delta}\right)\frac{u}{\Delta} = 0 \qquad (6.1)$$

with Δ, K, E as previously defined ((2.19), (2.29) and (2.30)).

We will make an approximation, which we call the low frequency approximation, which we will show *a posteriori* is in fact an approximation in $(M\omega/l)\to 0$, and is appropriate to the scattering problem: it is somehow 'connected' to infinity.

We proceed with the Schwarzschild case first; the scalar Kerr case will then be a straightforward extension. One trivially finds, by expanding in powers of r^{-1}, that (6.1) in Schwarzschild ($a=0$), is

$$\frac{d^2u}{dr^2}+\left(\omega^2+\frac{4M\omega^2}{r}-\frac{l(l+1)-12M^2\omega^2}{r^2}+O(r^{-3})\right)u=0. \quad (6.2)$$

Here all the terms involving ω^2 arise from the expansion of the combination K^2/Δ^2; E for Schwarzschild is just

$$E_l(\text{Schwarzschild})=l(l+1).$$

Because this is a large r approximation, it is a valid description of the scattering process if the wave function u is only appreciable when r is large. We can identify the implications of this requirement by computing the classical $r>0$ turning point associated with the multiplier of u in (6.2):

$$r_{TP}=2M\left[\left(\frac{l(l+1)}{(2M\omega)^2}-2\right)^{1/2}-1\right].$$

Clearly, for $(l/M\omega)\to\infty$, $r_{TP}\to\infty$, and the approximation (6.2) is the correct $\omega\to 0$ approximation, if $l\neq 0$. Notice also that the centrifugal barrier prevents absorption when $l\neq 0$. (Matzner (1968) has investigated the $l=0$ scalar case. He finds nonzero absorption, which leads to a logarithmically diverging $l=0$ partial scattering cross section: $\sigma_{l=0}\propto\ln\omega$.)

With the Schwarzschild situation in hand, we can proceed with the Kerr scalar case. Again we make an expansion in r^{-1} and neglect terms that fall off faster than r^{-3} at infinity. We find

$$\frac{d^2u}{dr^2}+\left(\omega^2+\frac{4M\omega^2}{r}-\frac{(E+a^2\omega^2-12M^2\omega^2)}{r^2}\right)u=0, \quad (6.3)$$

where the terms arise in essentially the same way as they did in the Schwarzschild expansion.

The scalar eigenvalue, E, in the Kerr case is (Breuer *et al.*, 1977 which corrects an error in Press & Teukolsky, 1973)

$$E_l=l(l+1)+2a^2\omega^2\left[\left(\frac{m^2-l(l+1)+\frac{1}{2}}{(2l-1)(2l+3)}\right)+O((a\omega)^4l^{-2})\right]. \quad (6.4)$$

Hence, for a fixed Kerr background, we can make the argument – valid for large enough l – that this large r approximation correctly describes the physics. When $(l/M\omega)$ is large, both the Kerr and the Schwarzschild radial equations then become

$$\frac{d^2u}{dr^2}+\left(\omega^2+\frac{4M\omega^2}{r}-\frac{l(l+1)}{r^2}\right)u=0; \tag{6.5}$$

this is the radial wave equation of the *comparison Newtonian problem*. (It can be recognized as the radial equation for the Schrödinger equation describing the attractive Coulomb problem with the identification of parameters:

$$-\frac{2Ze^2\mu}{\hbar^2}=4M\omega^2;\quad \frac{2\mu E}{\hbar^2}=\omega^2, \tag{6.6a}$$

where Ze^2 is the product of charges, μ is the mass of the scattered electron and E its energy; $\hbar$ is Planck's constant. Solving the first of these for M gives

$$-Ze^2/E=4M, \tag{6.6b}$$

see Merzbacher, 1971.) Hence we expect *Newtonian* gravity to be obeyed at this level of approximation. Notice that the appearance of ω^2 in the r^{-1} term guarantees that the equivalence principle is satisfied at this level.

We must tie up one loose end. Even though the Kerr radial equation becomes identical to the Schwarzschild one, we must verify that the angular equation becomes 'spherical' when l is large. This is in fact a well known result, since the angular function for $s=0$ is that for the spheroidal scalar harmonics ((2.28) and Flammer, 1957):

$$\frac{1}{\sin\theta}\frac{d}{d\theta}\left(\sin\theta\frac{dS}{d\theta}\right)+\left(a^2\omega^2\cos^2\theta-\frac{m^2}{\sin^2\theta}+E\right)S=0, \tag{6.7}$$

where S is the spheroidal harmonic. Since we deal with black holes, $a<M$; hence $l/(M\omega)\to\infty$ implies $l/(a\omega)\to\infty$. Then, because $E\to l(l+1)$ in the same limit, we see that (6.7) becomes that for the associated Legendre functions, i.e. the equation solved by the *spherical* scalar harmonics. The limit to the Newtonian case is then secure.

Hence, except for the anomalous $l=0$ case, the scattering is that given by the Rutherford formula:

$$\frac{d\sigma}{d\Omega}=\frac{M^2}{\sin^4(\theta/2)}. \tag{6.8}$$

We will discuss below the modification by the $l=0$ terms of the differential cross section (see also Matzner, 1968).

When the product $M\omega$ is small, the Newtonian form allows us to draw the following conclusion about the absorption: From (6.5), one finds that if

$M\omega \ll l$, then the classical turning point is $r_1 \sim l/\omega$. When $M\omega$ is small we can safely concentrate on the spherical (Schwarzschild) case. From (3.10), using the fact that the confluent hypergeometric function obeys (Gradshteyn & Ryzhik, 1965)

$$F(a, b, z) \underset{z\to 0}{\rightarrow} 1,$$

we see that $R_l \propto (\omega r)^l$ near $r = 0$, where

$$\phi = R_l P_l(\theta) e^{i\omega t}$$

in this limiting spherical case.

Can we consistently use this result in the black hole case? Yes we can, because the validity of the Newtonian approximation only requires $M\omega \ll l$. If $\omega M \ll \omega r_1 \ll l$, which can be in any case obtained by making ω small enough, then the non-Newtonian terms in (6.4) are negligible over all $r > r_1$. And the analysis just given shows that the wave function must be very small near the hole, except for $l = 0$, which is anomalous anyway. Thus the entire region outside r_1 is effectively Newtonian, and the wave function must remain in this Newtonian region.

For the comparison Newtonian problem the scattering cross section is obtained from the angular decomposition using (3.14). In particular, the phase shifts (denoted here η_l for the comparison Newtonian problem) are

$$e^{2i\eta_l} = \frac{\Gamma(l + 1 - 2iM\omega)}{\Gamma(l + 1 + 2iM\omega)}. \tag{6.9}$$

The decomposition (3.14) then gives the scalar version of the Newtonian scattering problem:

$$f(\theta) = \frac{(4\pi^{1/2})}{2i\omega} \sum_{l=0}^{\infty} (2l + 1)^{1/2} {}_0Y_l^0 (e^{2i\eta_l} - 1),$$

$$|f(\theta)|^2 = \frac{M^2}{\sin^4(\theta/2)}. \tag{6.10}$$

(In fact the $l = 0$ term in the relativistic case must be treated differently, see below; nonetheless this gives the *Newtonian* amplitude.)

Let us use the analysis of (3.1), (3.2) and (5.5) to calculate the absorption cross section of massless scalar waves in the low frequency limit (eventually limiting ourselves to Schwarzschild black holes). To that end, we can most efficiently proceed by using the scalar form of (2.65).

If

$$R = (r^2 + a^2)^{-1/2} \mathcal{Y}$$

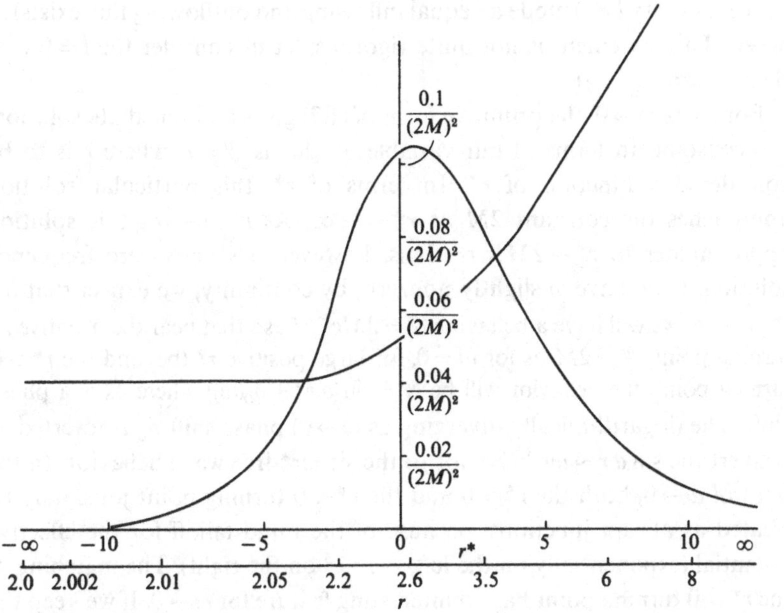

Fig. 6.1. The effective potential $V(r^*)$ for $l=0$ plotted against a scale linear in r^*. (r and r^* are in units of M). The constant in (1.5) has been set equal to zero. Also shown is the curve $u=r$ (not to scale) which is the solution to the s-wave radial equation for $\omega=0$.

then

$$0=\frac{d^2\mathcal{Y}}{dr^{*2}}+\left(\frac{[(r^2+a^2)\omega-am]^2-\lambda\Delta}{(r^2+a^2)^2}-\frac{r^2\Delta^2}{(r^2+a^2)^4}-\frac{d}{dr^*}\frac{r\Delta}{(r^2+a^2)^2}\right)\mathcal{Y}.$$

We present this equation for reference; its Schwarzschild form ($a\to 0$) is

$$0=\frac{d^2\mathcal{Y}}{dr^{*2}}+\left[\omega^2-\left(1-\frac{2M}{r}\right)\left(\frac{2M}{r^3}+\frac{l(l+1)}{r^2}\right)\right]\mathcal{Y}. \quad (6.11)$$

This form resembles a one-dimensional (second-order) Schrödinger equation. The effective potential $(1-2M/r)(2M/r^3)$ for the $l=0$ case is given in fig. 6.1. Because the relation between r and r^* is logarithmic, the potential term in this equation vanishes exponentially for large negative r^*. For large positive r^*, where $r\sim r^*$, the dominant feature of this equation when $l\neq 0$ is the centrifugal potential $\sim l(l+1)/r^{*2}$. Hence, for $\omega\to 0$ and nonzero l, the amplitude T of the wave $\mathcal{Y}$ near the horizon is very small compared with both R, that of the properly normalized scattered wave near infinity, and 1, the incident wave amplitude. Hence for $l\neq 0$, there is no absorption ($R^2=1$, hence for every $l \neq 0$ mode an equal inflowing and outflowing flux exists) as $\omega \to 0$. This statement is not quite rigorous; let us consider the $l=0$ case then return to $l \neq 0$.

For $l=0$, $\omega \to 0$, the primitive form of (1.2) gives an immediate solution: $\phi = \text{constant}$. In terms of our variable $\mathscr{Y}$, this is $\mathscr{Y} = r$, where r is to be considered a function of r^*. In terms of r^*, this particular solution approaches the constant $2M$ as $r^* \to -\infty$. As $r^* \to +\infty$, this solution approximates to $r^* - 2M \ln r^*$. It is, however, a strictly zero frequency solution. If we have ω slightly nonzero, by continuity, we expect that for $r^* \to -\infty$, we will have a behavior $\mathscr{Y} = 2M\mathrm{e}^{-i\omega r^*}$, so that near the negative r^* turning point, $\mathscr{Y} \simeq 2M$ as for $\omega = 0$; for large positive r^* (beyond the $r^* > 0$ turning point) the behavior will be $\mathscr{Y} = \sin(\omega r^* + \delta_0(\omega))$ where δ_0 is a phase shift. The (logarithmically diverging as $\omega \to 0$) phase shift δ_0 is inserted to convert the $\sin \omega r \sim \omega r$ behavior to the $\sin \omega r^*$ free wave behavior. In the limit of $\omega \to 0$, both the $r^* > 0$ and the $r^* < 0$ turning point joins may be treated as abrupt junctions, because of the rapid falloff for the effective potential (exponentially on the left, $\sim r^{*-3}$ on the right). The matching of the $r^* > 0$ turning point has an interesting feature for $\omega \to 0$. If we keep the normalization as above, the wave function which is $\sim 2M$ near the left turning point, is $\sim (1/2M\omega) \sin \omega r$ near the right turning point. This is matched to $a \sin(\omega r^* + \delta_0)$. But as $\omega \to 0$, $a \to 1/2M\omega$, because the matching process can introduce a different factor a only because the matching involves the wave function derivative. But $\mathrm{d}r/\mathrm{d}r^* \to 1$ as $r \to \infty$, and the matching point goes to infinity as $\omega \to 0$. Hence, rescaling the normalization, $\mathscr{Y} \sim (2M)^2$ near the left turning point has for its other asymptotic limit

$$\mathscr{Y} \sim (1/\omega) \sin(\omega r^* + \delta_0) \tag{6.12}$$

as $r^* \to \infty$. (The point of this final rescaling is that now the expression (6.12) has the same amplitude normalization as does the $l=0$ component of the incident 'plane' wave (3.16).)

The total energy flow down the hole due to a wave with $\mathscr{Y} = (2M)^2\mathrm{e}^{-i\omega r^*}\mathrm{e}^{-i\omega t}$ at the horizon is proportional to

$$\begin{aligned} |T^{0r}| &\propto \lim_{r \to 2M} |\omega \mathscr{Y}/r|^2 \cdot 4\pi r^2, \\ &= \omega^2 4\pi (2M)^2. \end{aligned} \tag{6.13}$$

The incident z-component flux (per unit area) of the 'plane' wave (3.16) at infinity is, with the same proportionality:

$$|T^{0z}| \propto \omega^2. \tag{6.14}$$

The $l=0$ absorption cross section in this limit, is the ratio of (6.13) to (6.14):

$$\sigma_{\text{abs}}(\omega \to 0) = 4\pi(2M)^2 \quad \text{(Schwarzschild).} \tag{6.15}$$

Caution: Matzner (1968) goes through much of this analysis, but deduces $\sigma_{\text{abs}} = 0$. Unruh (1976) was apparently the first to correctly calculate the result (6.15). He gives a much more technically complete treatment of this problem, including massive fields, and also treats the Dirac electron scattering from Schwarzschild black holes.

Notice that the amplitude near the horizon in the solution (6.12) is a factor $O(\omega)$ smaller than the amplitude near $r^* \to \infty$ (this is why Matzner (1968) incorrectly estimated the absorption) so we are correct in assuming that the wave near infinity has its elastic amplitude. Only a very small part of the $l=0$ wave falls into the hole as $\omega \to 0$. However, the $l=0$ part is strongly dominant in the decomposed wave as $\omega \to 0$. Hence the absorption cross section is nonzero and bounded, as given by (6.15).

The loose end we mentioned with regard to the $l \neq 0$ case concerns just how rapidly $\mathscr{Y}$ decreases to zero near the horizon in that case. After all, we have just seen that $\mathscr{Y} \sim r$ gives nonzero absorption. The falloff due to the centrifugal barrier is $\sim r^{l+1}$. The position of the outer turning point is $r \sim (l/\omega)$; this determines the place where falloff begins, hence the range over which the falloff occurs. Hence the amplitude near the horizon is $\sim (2M\omega/l)^{l+1}$ smaller than that at infinity; $l > 0$. The amplitude at infinity is $\sim \omega^{-1}$. Hence the flux at the horizon vanishes for $l \geqslant 1$ for $\omega \to \infty$. The absorption behavior has very interesting features as a function of frequency, for nonzero $M\omega$. These have been pointed out by Sanchez (1978 a, b). We defer discussion of her results to chapter 7 and chapter 8.

6.1.2 *Low frequency neutrino wave cross sections*

In the preceding example we saw that for the scalar case in the low frequency limit, one could safely assume a spherical Schwarzschild form, except for perhaps the *s*-wave portion of the angular decomposition, which piece might be absorbed into the black hole. For the purpose of obtaining and understanding low frequency results, we will assume that for the $s \neq 0$ case as well, the Schwarzschild solution contains the essential physics of the situation.

Then we may, for instance, consider the neutrino case. The radial wave equation for Schwarzschild can be obtained as an equation for $Z_{\pm} = r^{-1}(1-2M/r)^{-1/4} \; {}_{\pm 1/2}R_{lm\omega}$ in the form given by Chandrasekhar &

Detweiler (1977) (see also chapters 2 and 4):

$$0=\left[\frac{d^2}{dr^{*2}}+\omega^2-\frac{(l+\frac{1}{2})^2}{r^2}\left(1-\frac{2M}{r}\right)\mp\frac{l+\frac{1}{2}}{r^2}\left(1-\frac{2M}{r}\right)^{1/2}\left(1-\frac{3M}{r}\right)\right]Z_{\pm} \tag{6.16}$$

Here the two choices for the sign in the potential correspond to the two radial solutions $Z_{\pm}$. For this spherical case, Brill & Wheeler (1957) have shown that Z_+ and Z_- are the radial part of the upper and lower components of the neutrino 2-spinor (each multiplied by $r(1-2M/r)^{1/4}$). Also, r^* is the same as in the scalar case (1.5). The large-r limit of this equation is, for $u_{\pm}=Z_{\pm}(1-2M/r)^{1/2}$:

$$\frac{d^2u}{dr^2}+\left(\omega^2+\frac{4m\omega^2}{r}+\frac{12m^2\omega^2}{r^2}-\frac{k(k\pm 1)}{r^2}+O(r^{-3})\right)u_{\pm}=0 \tag{6.17}$$

where $k=l+\frac{1}{2}$ is a positive integer.

The exceptional case occurs when $k=1$; then there is apparently the possibility of the vanishing in the potential of the dominant contribution proportional to r^{-2}, and one component of the neutrino wave function can apparently 'leak' down to the horizon. For a discussion of that effect, we point out that Unruh (1976) has done a study of the *absorption* of neutrino waves on black holes, together with his study of the scalar absorption case.

Ignoring the problem of absorption, and taking $\omega\rightarrow 0$, the phase shifts associated with this scattering problem are, as in the scalar case (and the comparison Newtonian case, cf. (6.9))

$$e^{2i\delta_l^+}=\frac{\Gamma(1+k-2iM\omega)}{\Gamma(1+k+2iM\omega)}=\frac{\Gamma(1+l+\frac{1}{2}-2iM\omega)}{\Gamma(1+l+\frac{1}{2}+2iM\omega)}, \tag{6.18}$$

$$e^{2i\delta_l^-}=\frac{\Gamma(1+(k-1)-2iM\omega)}{\Gamma(1+(k-1)+2iM\omega)}=\frac{\Gamma(1+l-\frac{1}{2}-2iM\omega)}{\Gamma(1+l-\frac{1}{2}+2iM\omega)}. \tag{6.19}$$

These phase shifts, together with the angular behavior of the waves, can be combined into an expression for the differential cross sections:

$$d\sigma/d\Omega=|n^+(\theta)|^2, \tag{6.20}$$

where

$$n^+(\theta)=\frac{(4\pi)^{1/2}}{\omega}\sum_{j=1/2}^{\infty}(2j+1)^{1/2}\,{}_{-1/2}S_j^{1/2}(\theta)e^{2i\delta_j^+}. \tag{6.21}$$

The sum defining n^+ in (6.21) is complicated by the ubiquitous appearance of the number $\frac{1}{2}$. At present we know of no way to sum expression (6.21) for the low frequency neutrino scattering cross section.

This is doubly unfortunate because it will turn out that knowledge of the low frequency limit is essential in our *computation* of actual cross sections. The lack of such for the neutrinos prevents a computation of their cross section.

6.1.3 Low frequency limits of electromagnetic and gravitational scattering

For these two cases, analysis can be carried through completely. We find $\sigma_{\text{abs}} = 0$ and we obtain explicit closed form expressions for the $\omega \to 0$ scattering cross sections.

For gravitational and electromagnetic waves, the variables $K^{\text{up,down}}_{lm\omega P}$ described in (3.46), (3.50), (3.78) and (3.82) are simply related to the asymptotic form of the gauge independent wave variables (here collectively denoted X) described by Moncrief (1975). These wave variables X ($\sim e^{\pm i\omega r^*}$ at $r^* \to \pm\infty$) all satisfy a wave equation very similar to those for the scalar and the Dirac problem:

$$\left[\frac{\mathrm{d}^2}{\mathrm{d}r^{*2}} + \omega^2 - \left(1 - \frac{2M}{r}\right)\left(\frac{l(l+1)}{r^2} + V(r)\right)\right]X = 0 \tag{6.22}$$

where $V(r)$ is a function of order r^{-3}. ($V(r)$ is different for the electromagnetic, and the two ($P = \pm 1$) gravitational cases.) Further, for electromagnetism, $l \geqslant 1$ and for gravitational waves $l \geqslant 2$ in (6.22). Here we concentrate on the limit $\omega \to 0$.

The appearance of the term $l(l+1)/r^2$ in (6.22) shows that for $l \neq 0$, and for the $\omega \to 0$ limit considered here, the classical turning point is $r_{\text{TP}} \cong l/\omega$ for small ω. This makes the influence of the $(r)^{-3}$ terms of (6.22) negligible compared to the angular momentum term $l(l+1)/r^2$, and it means that the peak value of the potential barrier in (6.33) is very large compared to the energy, ω^2, of the wave. Hence the scattering becomes elastic for small $M\omega/l$, and a description in terms of (real) phase shifts becomes satisfactory. (We shall denote such phase shifts as γ_l.) The substitution

$$X = Y(1 - 2M/r)^{-1/2} \tag{6.23}$$

gives a wave equation of exactly the Newtonian comparison form, (6.5), for Y as $M\omega/l \to 0$. Further, because $l \geqslant 1$ here, we have *no absorption* for electromagnetic or gravitational waves as $M\omega \to 0$.

In the low frequency limit considered here, the two parities for gravitational wave scattering approach one another; a result which is straightforward from the form of the wave equation (6.5) in this limit. In general, for finite ω, the two gravitational wave parities scatter differently as we already noted in chapter 3. When dealing with the Riemann tensor formalism of chapter 3 the expression for the phase of the plane wave *is parity dependent*, with a parity dependence which disappears only at $\omega \to 0$ as considered here and at $\omega \to \infty$ (the WKB limit) (see section 6.2); this parity dependence of the plane wave then manifests itself in a parity dependence of the scattering in the intermediate cases.

We have seen that the phase shifts $\gamma_l^{(s)}$ tend to the Newtonian phase shifts η_l for $M\omega/l \to 0$ (here s labels the spin: 1 for electromagnetic and 2 for gravitational). Furthermore the two parities $P = \pm 1$ behave identically in this limit, even for scattering of gravitational waves. In the two cases, electromagnetic and gravitational, we can combine the phase shifts with the form of the plane wave mode constants to find a scattering amplitude for circularly polarized incident waves given in the $\omega \to 0$ limit. First, we write the scattering in terms of phase shifts, analogously to (3.14), (6.10) and (6.20):

Electromagnetic:

$$a(\theta) = \frac{(4\pi)^{1/2}}{2i\omega} \sum_{l=1}^{\infty} (2l+1)^{1/2} {}_{-1}S_l^1 (e^{2i\gamma_l^{(1)}} - 1), \tag{6.24a}$$

$$\underset{\omega \to 0}{\to} \frac{(4\pi)^{1/2}}{2i\omega} \sum_{l=1}^{\infty} (2l+1)^{1/2} {}_{-1}S_l^1 e^{2i\eta_l}. \tag{6.24b}$$

Gravitational:

$$g(\theta) = \frac{(4\pi)^{1/2}}{2i\omega} \sum_{l=2}^{\infty} (2l+1)^{1/2} {}_{-2}S_l^2 (e^{2i\gamma_l^{(2)}} - 1), \tag{6.25a}$$

$$\underset{\omega \to 0}{\to} \frac{(4\pi)^{1/2}}{2i\omega} \sum_{l=2}^{\infty} (2l+1)^{1/2} {}_{-2}S_l^2 e^{2i\eta_l}. \tag{6.25b}$$

In these two cases the second line, following the symbol $\to$, is obtained from the first by noticing that $\Sigma_{l=s}^{\infty} (2l+1)^{1/2} {}_s S_l^m$ is proportional to a δ-function in the forward direction. Since we anticipate the same forward divergence as in the Coulomb case, due to the long range Newtonian force, dropping this term which arises from the (-1) in (6.24a) and (6.25a) is a permissible transformation. The scattering amplitudes then are undefined up to a phase which amounts to adding a constant to each of the phase shifts, which in turn is equivalent to adjusting the constant in r^*, which is thus not relevant for calculating the cross section.

We use the formula (Gradshteyn & Ryzhik, 1965) with $x = \cos\theta$:

$$\int_{-1}^{1} [\sin^2(\theta/2)]^\sigma P_l(\cos\theta)\,\mathrm{d}\cos\theta$$

$$= \int_{-1}^{1} 2^{-\sigma}(1-x)^\sigma P_l(x)\,\mathrm{d}x = \frac{2(-1)^l\Gamma^2(\sigma+1)}{\Gamma(\sigma+l+2)\Gamma(1+\sigma-l)} \quad (\mathrm{Re}\,\sigma > -1) \tag{6.26}$$

This is the expression which is used to 'explain' the summation of the expression (6.10) for the scattering amplitude $f(\theta)$ for the scalar version of the comparison Newtonian problem (see also section 3.1, and Schiff (1968)). A small amount of manipulation using the properties of $\Gamma(z)$ gives

$$f(\theta) = \frac{(4\pi)^{1/2}}{2\mathrm{i}\omega}\sum_{l=0}^{\infty}(2l+1)^{1/2}\,{}_0S_l^0\,\mathrm{e}^{2\mathrm{i}\eta_l} \tag{6.27a}$$

$$= M[\sin^2(\theta/2)]^{-1+2\mathrm{i}M\omega}\mathrm{e}^{2\mathrm{i}\eta_0} \tag{6.27b}$$

and where (6.27a) is recognized as the scattering amplitude for the scalar limiting Newtonian problem. This explanation has to be taken with a small grain of salt, since the term $[\sin^2(\theta/2)]^{-1}$ in (6.27b) means that $\mathrm{Re}\,\sigma = -1$ in (6.26). However, we shall continue by viewing this integral as the limit as $\varepsilon \to 0_+$ for $\mathrm{Re}\,\sigma = -1 + \varepsilon$.

6.1.3.1 *Low frequency gravitational scattering*

The electromagnetic and gravitational cases now proceed in very similar ways. We present the (more complicated) gravitational case first, and then just sketch the electromagnetic case.

We introduce the raising operator L^+, which raises the z-component of angular momentum, m, by one unit* (Goldberg *et al.*, 1967)

$$L^+ = (\partial_\theta - m\cot\theta - s/\sin\theta). \tag{6.28}$$

We have

$$L^+\,{}_sS_l^m = [(l-m)(l+m+1)]^{1/2}\,{}_sS_l^{m+1}. \tag{6.29}$$

* The raising operator L^+ used here and defined by Goldberg *et al.* (1967) agrees for $s = 0$ with that used by Jackson (1962). Goldberg *et al.* (1967) also give an explicit form for ${}_sY_l^m$. This formula is only consistent with (6.29) and with Jackson's raising operator if an additional factor $(-1)^m$ is inserted in Goldberg *et al.* (1967), (3.1), and we make that change.

We also use the spin weight raising operator, $\eth$ (Goldberg *et al.*, 1967)

$$\eth = -(\partial_\theta - m/\sin\theta - s\cot\theta), \tag{6.30}$$

$$\eth\,{}_sS_l^m = [(l-s)(l+s+1)]^{1/2}\,{}_{s+1}S_l^m. \tag{6.31}$$

Now consider (6.27); since each term in the sum has $m=0$ and $s=0$, the operator L^+L^+ may be applied as a differential operator to the summed expression. We have

$$L^+L^+f(\theta) \equiv (\partial_\theta - \cot\theta)\partial_\theta f(\theta), \tag{6.32a}$$

$$= \frac{(4\pi)^{1/2}}{2i\omega}\sum_{l=0}^{\infty}(2l+1)^{1/2}L^+L^+\,{}_0S_l^0 e^{2i\eta_l}, \tag{6.32b}$$

$$= \frac{(4\pi)^{1/2}}{2i\omega}\sum_{l=0}^{\infty}(2l+1)^{1/2}[l(l-1)(l+1)(l+2)]^{1/2}\,{}_0S_l^2 e^{2i\eta_l}; \tag{6.32c}$$

a sum which is even more divergent than (6.27b), but which may be evaluated by explicitly applying the differential operators to the expression $f(\theta)$ of (6.27b). Notice that the $l=0$, $l=1$ terms of (6.32c) vanish.

Similarly, since each term in the sum (6.25b) has the same values $s=-2$, $m=2$, we have

$$\eth\eth g(\theta) = \left(\partial_\theta - \frac{2}{\sin\theta} + \cot\theta\right)\left(\partial_\theta - \frac{2}{\sin\theta} + 2\cot\theta\right)g(\theta), \tag{6.33a}$$

$$= \frac{(4\pi)^{1/2}}{2i\omega}\sum_{l=2}^{\infty}(2l+1)^{1/2}\eth\eth\,{}_{-2}S_l^2 e^{2i\eta_l} \tag{6.33b}$$

$$= \frac{(4\pi)^{1/2}}{2i\omega}\sum_{l=2}^{\infty}(2l+1)^{1/2}[l(l-1)(l+1)(l+2)]^{1/2}\,{}_0S_l^2 e^{2i\eta_l}. \tag{6.33c}$$

The result (6.33c) is a sum identical to that of (6.32c). We shall not concern ourselves with the divergent nature of these sums, since they obviously sum to the quantity

$$H = \sin\theta\,\partial_\theta\left(\frac{\partial_\theta f}{\sin\theta}\right) \tag{6.34}$$

which diverges at $\theta=0$. Instead we regard (6.32c) and (6.33c) as giving us a linear second-order inhomogeneous differential equation to solve for $g(\theta)$:

$$\left(\partial_\theta - \frac{2}{\sin\theta} + \cot\theta\right)\left(\partial_\theta - \frac{2}{\sin\theta} + 2\cot\theta\right)g(\theta) = H. \tag{6.35}$$

Since the second-order operator is presented in a factored form, the integration of (6.35) is straightforward. Using the notation $y = \sin^2(\frac{1}{2}\theta)$ we obtain

$$g(\theta) = \frac{B}{\cos^4(\frac{1}{2}\theta)} + \frac{\tilde{B}y}{\cos^4(\frac{1}{2}\theta)} + g_{\text{I}}, \tag{6.36}$$

$$\cos^4(\tfrac{1}{2}\theta)g_{\text{I}} = Me^{2i\eta_0}\left\{y^{-1+2iM\omega} + \left(\frac{2iM\omega - 2}{2iM\omega}\right)\right.$$
$$\left.\times\left[\left(\frac{2iM\omega - 1}{2iM\omega + 1}\right)y^{1+2iM\omega} - 2y^{2iM\omega}\right]\right\}, \tag{6.37}$$

where B and $\tilde{B}$ are the two constants of integration associated with the two solutions to the homogeneous version of (6.35) and g_{I} is the inhomogeneous solution. The constants B and $\tilde{B}$ can be determined by obtaining the moment of (6.36) against any two convenient ${}_{-2}S_l^2$. Since

$$2\pi\int {}_sS_l^m\,{}_sS_{l'}^m\,\mathrm{d}x = \delta_{ll'}, \quad x = \cos\theta, \tag{6.38}$$

we have

$$\frac{2}{2i\omega}\left(\frac{2l+1}{4\pi}\right)^{1/2} e^{2i\eta_l} = \int g(\theta)\,{}_{-2}S_l^2\,\mathrm{d}x$$
$$= B\int_{-1}^{1} {}_{-2}S_l^2\frac{\mathrm{d}x}{\cos^4(\frac{1}{2}\theta)} + \tilde{B}\int_{-1}^{1} {}_{-2}S_l^2\frac{y}{\cos^4(\frac{1}{2}\theta)}\mathrm{d}x$$
$$+ \int_{-1}^{1} g_{\text{I}}\,{}_{-2}S_l^2\,\mathrm{d}x \tag{6.39}$$

so evaluation of the integrals for any two l values uniquely determines B and $\tilde{B}$. Because the polynomials' expressions become more complicated for larger l, the sensible procedure is to use ${}_{-2}S_2^2$ and ${}_{-2}S_3^2$ to determine the values of B and $\tilde{B}$.

Now (Goldberg *et al.* (1967); see footnote on p. 87):

$${}_{-2}S_2^2 = \left(\frac{5}{4\pi}\right)^{1/2}\cos^4(\tfrac{1}{2}\theta) \tag{6.40}$$

while

$${}_{-2}S_3^2 = \left(\frac{7}{4\pi}\right)^{1/2}\cos^4(\tfrac{1}{2}\theta)[1 - 6\sin^2(\tfrac{1}{2}\theta)]. \tag{6.41}$$

The integrals are tedious but straightforward. We obtain

$$B = -\frac{2e^{2i\eta_0}}{2i\omega}\left(\frac{2 - 2iM\omega}{1 + 2iM\omega}\right), \tag{6.42}$$

$$\tilde{B} = -\frac{2e^{2i\eta_0}}{2i\omega}; \tag{6.43}$$

which gives

$$g(\theta) = Me^{2i\eta_0}\frac{y^{2iM\omega}}{y} + \frac{2ye^{2i\eta_0}}{2i\omega}\left(y^{2iM\omega}\frac{1-4iM\omega}{1+2iM\omega} - 1\right)[\cos^4(\tfrac{1}{2}\theta)]^{-1}$$

$$+\frac{4e^{2i\eta_0}}{2i\omega}\left(y^{2iM\omega} - \frac{1-iM\omega}{1+2iM\omega}\right)[\cos^4(\tfrac{1}{2}\theta)]^{-1}. \quad (6.44)$$

The first term in this sum is identical to the $f(\theta)$ for the comparison scalar Newtonian problem.

The last two terms in the expression (6.44) have an apparent divergence in the backward direction, $\theta \to \pi$. However, this divergence is not real, as we now show. Now

$$y^{2iM\omega} \equiv \exp\{2iM\omega \ln[1 - \cos^2(\tfrac{1}{2}\theta)]\}. \quad (6.45)$$

As $\cos^2(\tfrac{1}{2}\theta) \to 0$,

$$y^{2iM\omega} = 1 - 2iM\omega\cos^2(\tfrac{1}{2}\theta) + \tfrac{1}{2}\cos^4(\tfrac{1}{2}\theta)(2iM\omega)(2iM\omega - 1) + O(\cos^6(\tfrac{1}{2}\theta)). \quad (6.46)$$

In this limit, the trigonometric factors in the last two terms combine to cancel the $[\cos^4(\tfrac{1}{2}\theta)]^{-1}$ denominator. We find

$$g(\theta \to \pi) \to Me^{2i\eta_0}[1/y - 1 + O(\cos^2(\tfrac{1}{2}\theta))] \quad (6.47)$$

which vanishes at $\theta = \pi$. There is thus a remarkable cancellation of the backward divergence arising from the homogeneous terms B and $\tilde{B}$ in (6.36). In fact, as fig. 6.2 shows (calculated from (6.44) for $M\omega = 0.1$), the amplitude is remarkably small in the backward direction.

The Newtonian problem calculated here approaches the relativistic scattering problem when the frequency becomes small, $M\omega \to 0$. The round brackets in the last two terms in (6.44) contain the difference of quantities equal to $1 + O(M\omega)$ as $M\omega \to 0$ so the *a priori* low frequency divergence in the last two terms is avoided: As $M\omega \to 0$*,

$$y^{2iM\omega} \equiv \exp[2iM\omega \ln \sin^2(\tfrac{1}{2}\theta)] \quad (6.48)$$

$$\cong 1 + 2iM\omega \ln \sin^2(\tfrac{1}{2}\theta). \quad (6.49)$$

* Except for $\theta \to 0$, where (6.49) fails. For sufficiently small ω, the oscillatory behavior of (6.48) is squeezed into a forward cone of size $\theta \sim \exp(-1/4M\omega)$, and as $\omega \to 0$ is lost in the forward peak of the amplitude.

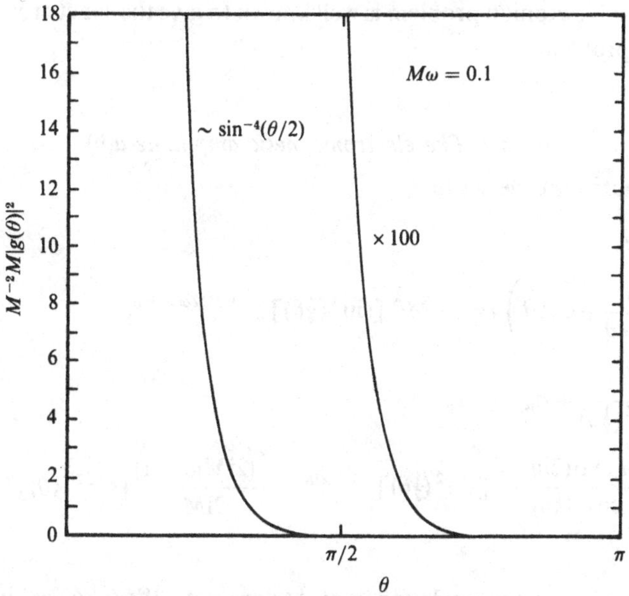

Fig. 6.2. The low frequency limit of gravitational radiation scattering on a Schwarzschild black hole, calculated via (6.44) for $M\omega = 0.1$, see Matzner & Ryan (1977). As shown in (6.47), $g(\theta)$ vanishes for $\theta \to \pi$, $M\omega \to 0$. This analytic form will be used to 'sum' the partial wave expansion to obtain the cross sections reported in figs. 8.13–8.17 and 8.19–6.20. This is done by replacing the $l \leqslant 20$ amplitudes in the limiting form by the numerically calculated values (see chapter 7). For sufficiently small ω/l, the partial wave sum calculated there converges to the analytic result.

Hence, in the relevant limit $M\omega \to 0$

$$g(\theta) \to \mathrm{M}e^{2i\eta_0} y^{2iM\omega}/y$$

$$+ \frac{2M \exp(2i\eta_0)}{\cos^4(\frac{1}{2}\theta)} \{y[\ln \sin^2(\tfrac{1}{2}\theta) - 3] + 2\ln \sin^2(\tfrac{1}{2}\theta) + 3\}. \quad (6.50)$$

Although qualitatively similar, the cross section calculated from the result (6.50) differs in detail from that found for the Born approximation by Westervelt (1971) and Peters (1976):

$$\sigma(\theta) = M^2 \frac{\sin^8(\frac{1}{2}\theta) + \cos^8(\frac{1}{2}\theta)}{\sin^4(\frac{1}{2}\theta)}. \quad (6.51)$$

This difference is even more remarkable since the Born result for the scalar case of the Newtonian problem is well known to give the exact cross section for that problem.

6.1.3.2 The electromagnetic amplitude $a(\theta)$

To evaluate $a(\theta)$, we write

$$\eth a = L^+ f, \tag{6.52}$$

$$\left(\partial_\theta - \frac{1}{\sin\theta} + \cot\theta\right)a = -M\partial_\theta[\sin^2(\tfrac{1}{2}\theta)]^{-1+2iM\omega}e^{2i\eta_0}, \tag{6.53}$$

$$a = \frac{\tilde{A}}{\cos^2(\tfrac{1}{2}\theta)} + a_I, \tag{6.54}$$

$$a_I = -\frac{M\exp(2i\eta_0)}{\cos^2(\tfrac{1}{2}\theta)}\left([\sin^2(\tfrac{1}{2}\theta)]^{-1+2iM\omega} - \frac{(2iM\omega-1)}{2iM\omega}[\sin^2(\tfrac{1}{2}\theta)]^{2iM\omega}\right). \tag{6.55}$$

Here there is only the single constant $\tilde{A}$ to evaluate. We take, for simplicity, the moment of a against ${}_{-1}S_1^1$ (Goldberg *et al.* (1967); n.b. footnote on p. 87. Now

$${}_{-1}S_1^1 = -\left(\frac{3}{4\pi}\right)^{1/2}\cos^2(\tfrac{1}{2}\theta). \tag{6.56}$$

Hence

$$\int a\,{}_{-1}S_1^1\,dx = \frac{(12\pi)^{1/2}e^{2i\eta_1}}{2\pi 2i\omega} = 2\left(\frac{3}{4\pi}\right)^{1/2}\tilde{A} + \int a_I\,{}_{-1}S_1^1\,dx. \tag{6.57}$$

We obtain

$$\tilde{A} = (2i\omega)^{-1}e^{2i\eta_0} \tag{6.58}$$

which gives

$$e^{-2i\eta_0}a\cos^2(\tfrac{1}{2}\theta) = -M\{[\sin^2(\tfrac{1}{2}\theta)]^{-1+2iM\omega} - [\sin^2(\tfrac{1}{2}\theta)]^{2iM\omega}\} - \frac{1}{2i\omega}\{[\sin^2(\tfrac{1}{2}\theta)]^{2iM\omega} - 1\}. \tag{6.59}$$

Using (6.48) and (6.49) for small $M\omega$, the scattering amplitude except near the forward direction is

$$a = -Me^{2i\eta_0}\left(\frac{[\sin^2(\tfrac{1}{2}\theta)]^{2iM\omega}}{\sin^2(\tfrac{1}{2}\theta)} + \frac{\ln\sin^2(\tfrac{1}{2}\theta)}{\cos^2(\tfrac{1}{2}\theta)}\right). \tag{6.60}$$

Finally, we may investigate the $\theta \to \pi$ (backward) limit of this result using (6.46)

$$a \underset{\theta \to \pi}{\to} -Me^{2i\eta_0}\left(\frac{1}{\sin^2(\frac{1}{2}\theta)} - 1\right),$$

$$\cong -Me^{2i\eta_0}\left(\frac{\cos^2(\frac{1}{2}\theta)}{\sin^2(\frac{1}{2}\theta)}\right). \tag{6.61}$$

Only in this limit does the scattering amplitude agree with the form

$$\sigma(\theta) = M^2\left(\frac{\cos^4(\frac{1}{2}\theta)}{\sin^4(\frac{1}{2}\theta)}\right) \tag{6.62}$$

found by Westervelt (1971) and Peters (1976) in Born approximation calculations.

6.2 High frequency cross sections

In the limit $M\omega \to \infty$, one expects wave behavior to go over to particle motion, following the usual eikonal (JWKB: Jeffries, 1923; Wentzel, 1926; Kramers, 1926; Brillouin, 1926) expansion of the phase.

6.2.1 Eikonal approximation for integer spin fields

The behavior of massless scalar waves is governed by the equation

$$(\phi_{,\mu})^{;\mu} = 0. \tag{6.63}$$

We will suppose that $\phi = \Phi e^{i\psi}$ where Φ, ψ are functions of position so chosen that $|\partial\psi/\psi| \gg |\partial\Phi/\Phi|$ (i.e., every logarithmic derivative of Φ is much smaller than any logarithmic derivative of ψ). Compute

$$\phi_{,\mu} = \Phi_{,\mu}e^{i\psi} + i\phi\psi_{,\mu} \tag{6.64}$$

and

$$\Box\phi = \Phi_{,\mu}{}^{;\mu}e^{i\psi} + 2i\Phi^{;\mu}\psi_{,\mu}e^{i\psi} - \phi\psi_{,\mu}\psi^{;\mu} + i\phi\psi_{,\mu}{}^{;\mu}. \tag{6.65}$$

In the eikonal approximation, we regard the terms involving $\psi_{,\mu}$ to be dominant, and slowly changing, so the leading term of (6.63) is

$$0 = \phi\psi_{,\mu}\psi^{;\mu}. \tag{6.66}$$

This shows that the vector $k_\mu = \psi_{,\mu}$ is *null*. It also shows that k_μ is geodesic:

$$0 = (g^{\alpha\beta}\psi_{,\alpha}\psi_{,\beta})_{;\sigma} = 2g^{\alpha\beta}\psi_{,\alpha;\sigma}\psi_{\beta}, \tag{6.67}$$

$$0 = 2\psi^{\alpha}\psi_{,\alpha;\sigma}. \tag{6.68}$$

However it is straightforward to verify that

$$\psi_{,\alpha;\sigma} = \psi_{,\sigma;\alpha}. \tag{6.69}$$

Hence

$$\psi^{\alpha}\psi_{,\sigma;\alpha} = k^{\alpha}k_{\sigma;\alpha} = 0, \tag{6.70}$$

which shows that k_μ is tangent to an affinely parameterized geodesic.

The next order in this expansion then contains

$$2\Phi_{,\mu}k^{\mu} + \Phi k^{\mu}{}_{;\mu} = 0, \tag{6.71}$$

or, with $k^\mu \partial_\mu \equiv \mathrm{d}/\mathrm{d}\lambda$:

$$\frac{\mathrm{d}}{\mathrm{d}\lambda}\Phi^2 + \Phi^2(\text{divergence } k) = 0, \tag{6.72}$$

which expresses the fact that the amplitude of the wave decreases along its path as the rays diverge.

For electromagnetic and gravitational perturbations the calculation is similar. We consider the electromagnetic potential A^μ, and the gravitational metric perturbation $h_{\mu\nu}$, each on a fixed (curved) background. The equations obeyed by each (in vacuum) are

$$A^{\mu}{}_{;\sigma}{}^{;\sigma} = 0, \tag{6.73}$$

$$A^{\mu}{}_{;\mu} = 0, \tag{6.74}$$

$$h_{\mu\nu;\alpha}{}^{;\alpha} + 2R^{(\mathrm{B})}_{\alpha\mu\beta\nu}h^{\alpha\beta} = 0, \tag{6.75}$$

$$h_{\mu}{}^{\mu} = 0 = h_{\mu\beta}{}^{;\beta}; \tag{6.76}$$

where the covariant derivative in (6.73)–(6.76) is taken using the background metric form, and $R^{(\mathrm{B})}_{\alpha\mu\beta\nu}$ is the background Riemann tensor.

In the electromagnetic case one has the simple wave operator equation for A^μ only if the Lorentz gauge (6.74) is obeyed; for gravitation one requires (6.76): vanishing trace $h^{\alpha}{}_{\alpha}$, and a Lorentz-like condition $h_{\alpha\beta}{}^{;\beta} = 0$ (see Misner *et al.*, 1973, chapter 35).

The analysis then can be carried out in parallel: Suppose $A_\mu = \mathscr{A}_\mu e^{i\psi}$, $h_{\mu\nu} = \mathscr{H}_{\mu\nu}e^{i\psi}$. Then

Gauge condition:

$$\mathscr{H}_{\mu}{}^{\mu} = 0; \quad \mathscr{H}_{\mu}{}^{\nu}{}_{;\nu} + i\mathscr{H}_{\mu}{}^{\nu}\psi_{;\nu} = 0, \tag{6.77}$$

$$\mathscr{A}^{\mu}{}_{;\mu} + i\mathscr{A}^{\mu}\psi_{;\mu} = 0. \tag{6.78}$$

Equations of motion:

$$0 = i\mathscr{H}_{\mu\nu}\psi_{,\alpha}{}^{;\alpha} + 2\mathscr{H}_{\mu\nu}{}^{;\alpha}i\psi_{;\alpha} + \mathscr{H}_{\mu\nu;\alpha}{}^{;\alpha} - \psi_{,\alpha}\psi^{;\alpha}\mathscr{H}_{\mu\nu} + 2R^{(\mathrm{B})}{}_{\alpha\mu\beta\nu}\mathscr{H}^{\alpha\beta}, \tag{6.79}$$

$$0 = \mathscr{A}^{\mu}{}_{;\sigma}{}^{;\sigma} + 2i\mathscr{A}^{\mu}{}_{;\sigma}\psi^{;\sigma} + i\mathscr{A}^{\mu}\psi_{;\sigma}{}^{;\sigma} - \mathscr{A}^{\mu}\psi_{,\sigma}\psi^{;\sigma}. \quad (6.80)$$

As before we consider $\psi_{,\mu}$ large, while $\psi\psi_{,\sigma}{}^{;\beta}/\psi_{;\alpha}\psi_{;\mu}$ is small, and derivatives of $\mathscr{A}^{\mu}$ and of $\mathscr{H}^{\mu\nu}$ are small compared to those of ψ. (We use the background metric to raise and lower indices.) Also $|\psi_{,\sigma}|^2 \gg |R^{(B)}{}_{\alpha\beta\gamma\delta}|$. The dominant terms in these equations are then

Gauge conditions (dominant order):

$$\mathscr{H}_{\mu}{}^{\mu} = 0; \quad \mathscr{H}_{\mu}{}^{\nu}\psi_{,\nu} = 0, \quad \mathscr{A}^{\mu}\psi_{,\mu} = 0. \quad (6.81)$$

Equations of motion:

$$-\psi_{,\alpha}\psi^{;\alpha}\mathscr{H}_{\mu\nu} = 0, \quad -\psi_{,\alpha}\psi^{;\alpha}\mathscr{A}^{\mu} = 0. \quad (6.82)$$

Here the equations of motion show that, as before, $\psi_{,\alpha} = k_{\alpha}$ is a null vector; it then follows as before that k_{α} is geodesic, cf (6.70). The gauge conditions then state that $\mathscr{H}_{\alpha\beta}$ is traceless, and both A^{μ} and $h_{\mu\nu}$ are *transverse.*

The subdominant parts of the equations of motion are

$$\mathscr{H}_{\mu\nu}\psi_{;\alpha}{}^{;\alpha} + 2\mathscr{H}_{\mu\nu;\alpha}k^{\alpha} = 0, \quad (6.83)$$

$$\mathscr{A}^{\mu}k_{\alpha}{}^{;\alpha} + 2\mathscr{A}_{;\alpha}{}^{\mu}k^{\alpha} = 0. \quad (6.84)$$

If we introduce spacelike polarization tensors:

$$\mathscr{H}_{\alpha\beta} = \mathscr{H}e_{\alpha\beta}, \quad (6.85)$$

$$\mathscr{A}_{\mu} = \mathscr{A}e_{\mu}, \quad (6.86)$$

then $k^{\mu}e_{\mu\alpha} = k^{\nu}e_{\nu} = 0$, $e^{\alpha\beta}e_{\alpha\beta} = e^{\alpha}e_{\alpha} = 1$ and equations (6.83), (6.84) can be rewritten to read

$$\mathscr{H}^2k^{\alpha}{}_{;\alpha} + (\mathscr{H}^2)_{,\alpha}k^{\alpha} = 0; \quad e_{\mu\nu;\alpha}k^{\alpha} = 0, \quad (6.87)$$

$$\mathscr{A}^2k^{\alpha}{}_{;\alpha} + (\mathscr{A}^2)_{,\alpha}k^{\alpha} = 0; \quad e_{\mu;\alpha}k^{\alpha} = 0. \quad (6.88)$$

As in the scalar case, these show the evolution of the wave strength, decreasing as k diverges. In addition, we have the result that the polarization is parallelly transported along the orbit: $e_{\mu\nu;\alpha}k^{\alpha} = e_{\mu;\alpha}k^{\alpha} = 0$.

These equations embody in purest form the high frequency limit. Radiation follows null geodesics, and carries its polarization parallely along with it.

6.2.2 Eikonal approximation for neutrino fields

A discussion for neutrinos, similar to the one in the preceding section, is complicated by the fact that one must work in a tetrad frame. The neutrino equation in such a frame is

$$\nabla_{A\dot{A}}\zeta^A = 0 \tag{6.89}$$

where

$$\nabla_{A\dot{A}} \equiv \sigma^{\mu}_{A\dot{A}}\nabla_{\mu}, \tag{6.90}$$

σ^{μ}_{AA} is an Infeld–van der Waerden symbol and must be specified (cf sections 4.2–4.5) for an explicit definition, and ∇_{μ} here is the spinor covariant derivative defined in (4.17) and (4.18).

The choice of the basis can obviously affect the value of the spinor covariant derivative. However, we now assume that ζ^A can be expressed in the JWKB form

$$\zeta^A = S^A e^{i\psi}. \tag{6.91}$$

We then have

$$\nabla_{A\dot{A}}\zeta^A = 0 = (\nabla_{A\dot{A}}S^A)e^{i\psi} + \psi_{,A\dot{A}}S^A e^{i\psi} = 0. \tag{6.92}$$

We can then repeat this process:

$$\nabla^{B\dot{A}}\nabla_{A\dot{A}}\zeta^A = 0. \tag{6.93}$$

If the geometry in which the neutrino propagates is fairly smooth, and the basis tetrad properly chosen, then it is appropriate to suppose that the derivatives of ψ will dominate this equation. Thus, to dominant order:

$$\nabla^{B\dot{A}}\nabla_{A\dot{A}}\zeta^A \cong \psi^{,B\dot{A}}\psi_{,A\dot{A}}S^A = 0. \tag{6.94}$$

Written out using the van der Waerden matrices, this becomes

$$\sigma_{\mu}{}^{B\dot{A}}\sigma^{\nu}{}_{A\dot{A}}\psi^{,\mu}\psi_{,\nu}S^A = 0. \tag{6.95}$$

In our tetrad, we will take the $\sigma^j_{A\dot{B}}$ to have the values given by (4.32), (4.33), in which case one finds

$$\sigma_{\mu}{}^{B\dot{A}}\sigma^{\nu}{}_{A\dot{A}} = \tfrac{1}{2}\delta_{\mu}{}^{\nu}\varepsilon^{B}{}_{A}. \tag{6.96}$$

Because of the invariant form of the right-hand side of this equation, it must hold in all spin frames and in all coordinate frames. We immediately have – as before for the integral spin cases – that $\psi_{,\mu} = k_{\mu}$ is null, and geodesic.

One of the consequences of (4.4) connecting the spacetime and spin metrics is that if $\psi_{,\mu} = \sigma^{A\dot{A}}_{\mu}\psi_{,A\dot{A}}$ is null, then

$$\psi_{,A\dot{A}} = \psi_A\bar{\psi}_{\dot{A}}, \tag{6.97}$$

for some spinor ψ_A. Thus $\psi_{,AA}$ is a simple product of spinors, which expression is possible only because $\psi_{,\mu}$ is null (NP). Consequently the

JWKB neutrino equation (comprising the dominant terms in (6.92)) reads

$$\psi_{,A\dot{A}}S^A = 0, \tag{6.98}$$

or

$$\bar{\psi}_{\dot{A}}\psi_{,A}S^A = 0. \tag{6.99}$$

If $\psi_{,A}S^A = \varepsilon^{AB}\psi_{,A}S_A = 0$, S_A and $\psi_{,A}$ are proportional. We thus have that the helicity $\sigma^{\mu}_{A\dot{A}}S^A\bar{S}^{\dot{A}} = \alpha S^{\mu} = k^{\mu}$ (with α a perhaps negative but nonzero factor). Thus the helicity S is parallelly propagated because, by the geodesic equation, k_{μ} is. S^{μ} is either parallel, or antiparallel, to k_{μ}. See Choquet-Bruhat (1969) for an alternative discussion of the parallel transport of polarization.

6.2.3 On-axis observers: glories in high frequency scattering

For all spins investigated, then, the high-frequency limit of massless motion is to follow geodesics. We anticipate that this will be appropriate in most situations, at least for those that avoid singularities of the null-ray field. For instance, the Einstein estimate of the deflection angle for photon trajectories passing the sun:

$$\theta_{\text{defl}} = 4M/b \quad (\theta_{\text{defl}} \text{ small}) \tag{6.100}$$

(b is the impact parameter) leads to a cross section (cf Goldstein, 1950)

$$\frac{d\sigma}{d\Omega} = \left|\frac{1}{2}\frac{db^2}{d\cos\theta}\right| \sim M^2 \bigg/ \left(\frac{\theta}{2}\right)^4 \tag{6.101}$$

which is the small angle approximation of (6.8).

The classical behavior of the forward divergence, associated with the concept of gravitational lensing, can be investigated by a technique introduced by Einstein (1936).

The ray optics situation cannot be extracted directly from the cross section, since that is evaluated strictly at infinity, but we can run through the analysis with the aid of fig. 6.3.

The observer is at a distance R from the scatterer. The ray from the infinitely distant source (impact parameter b) intersects the axis at R_b. A ray with slightly bigger b intersects the axis at R_{b+db}. The observer at R thus sees the rays which were initially in the annulus with area $2\pi b db$ to lie now in the annulus of area $2\pi D dD$. We use the Einstein angle: $\theta = 4M/b$.

Now

$$D = \theta(R - R_b) = 4MR/b - R_b\theta, \tag{6.102}$$

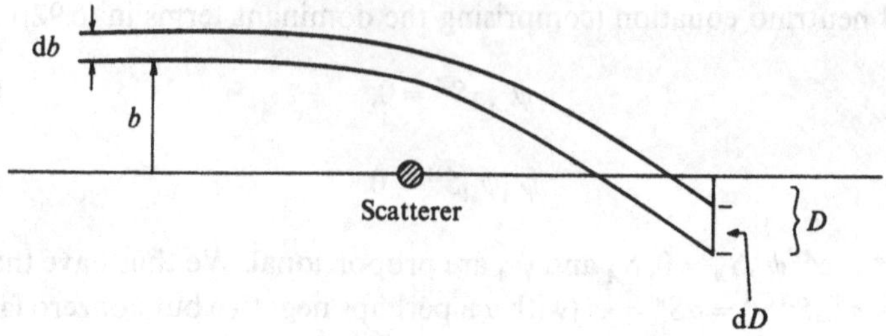

Fig. 6.3. The scatterer causes caustic focusing of incident rays onto the axis. An observer at distance R down the axis, D off-axis, observes rays with initial impact parameter b.

but the crossing point R_b is defined by

$$R_b\theta = b, \tag{6.103}$$

so

$$D = 4MR/b - b. \tag{6.104}$$

Hence

$$\left|\frac{D\,\mathrm{d}D}{b\,\mathrm{d}b}\right| = \left(\frac{4MR}{b^2} + 1\right)\frac{D}{b}. \tag{6.105}$$

Because $R_b \cong R$ (so $(4M/b)(R/b) = \theta R/b \simeq \theta R_b/b \simeq 1$), this is

$$\frac{D\,\mathrm{d}D}{b\,\mathrm{d}b} \cong 2\frac{D}{b}. \tag{6.106}$$

Further, $R_b(4M/b) = R_b\theta = b$ also gives $b = (4MR_b)^{1/2}$, so the amplification factor is

$$(\text{ratio of areas})^{-1} \sim 2(MR)^{1/2}/D, \tag{6.107}$$

where the explicit multiplier 2 comes from the fact that a similar amplification occurs for rays that cross the axis a distance $|R - R_b|$ behind the observer. Note that this gives divergent on-axis behavior.

Equations (6.101) and (6.107) were, however, derived under the assumption that radiation is accurately described by null geodesics. This analysis fails precisely where (6.101) and (6.107) diverge, i.e. on the axis, because there the different null rays intersect, so there one would expect interference between the different wave components.

We will first give an analysis for scalar massless waves. The propagation of polarization is of course an important separate problem which we take up below based on the suggestive analysis of the scalar situation.

The following analysis of the on-axis behavior in the weak-field, high frequency limit is due to Jackson (1985). We consider the on-axis behavior, far from the scatterer.

The divergence on the axis comes from large-l components in the partial-wave sum of solutions to (6.1). Hence one may safely take the limiting Newtonian form of this equation, i.e. (6.5), to discuss the on-axis scattering. Then we are dealing with the partial wave expansion of

$$\nabla^2\phi + (\omega^2 + 4\omega^2 M/r)\phi = 0. \tag{6.108}$$

This is Jackson's fundamental equation. It is, not surprisingly, an equation for Newtonian scattering, identical to the three-dimensional form of the Newtonian scattering problem, when the identifications (6.6), are made.

We have already given the spherical coordinate partial wave solution to this equation (cf (3.9), (3.10), (6.9), (6.10)). However, in cylindrical coordinates, z, ρ, ϕ it is possible to write the exact scattering solution (wave incident from the negative z direction):

$$\phi = \mathrm{e}^{\pi\omega M}\Gamma(1 - \mathrm{i}\omega M)\mathrm{e}^{\mathrm{i}\omega z}{}_1F_1(\mathrm{i}\omega M, 1; \mathrm{i}\omega(r - z)) \tag{6.109}$$

(where ${}_1F_1$ is a confluent hypergeometric function and $r^2 = \rho^2 + z^2$). Take $\omega > 0$ and consider the behavior of ${}_1F_1$ for large negative z:

$$\phi \to \mathrm{e}^{\mathrm{i}\omega z}. \tag{6.110}$$

Evaluating $\Gamma(1 - \mathrm{i}\omega M)\Gamma(1 + \mathrm{i}\omega M) = \pi M\omega/\sinh(\pi M\omega)$, the scattered amplitude near the axis ($0 < \rho \ll z$) gives

$$\text{gain} = \phi\bar{\phi} = \frac{2\pi\omega M \mathrm{e}^{2\pi\omega M}}{\sinh^2 \pi\omega M}|{}_1F_1|^2 \tag{6.111}$$

when $r = z$, ${}_1F_1 = 2^{-1/2}$, which gives the gain in intensity over the incident plane wave. Clearly, as $\omega M \to \infty$, the gain on axis is

$$4\pi\omega M. \tag{6.112}$$

Now, there is a standard limiting form for the confluent hypergeometric function:

$$\lim_{a\to\infty} {}_1F_1(a, c, -y) = \Gamma(c)(ay)^{1/2 - c/2} J_{c-1}((2ay)^{1/2}), \tag{6.113}$$

which means that our function ${}_1F_1(\mathrm{i}\omega M, 1, \mathrm{i}\omega r(1 - \cos\theta))$ has the limit

$$J_0(2\sqrt{2}M^{1/2}\omega R^{1/2}(1 - \cos\theta)^{1/2}). \tag{6.114}$$

Now, near the forward direction, $R(1 - \cos\theta) \approx R\theta^2/2 \approx D^2/2R$, where D is, as above, the distance off the axis. Further, when $(M\omega^2)D^2/2R$ is large, i.e.

when D is somewhat off the axis, we can use the asymptotic form of J_0:

$$J_0(x) \sim \left(\frac{2}{\pi x}\right)^{1/2} \times \text{oscillatory terms.} \tag{6.115}$$

Then, when $M\omega^2 D^2/2R \gg 1$, one has the average gain, averaged over one of the oscillations of the Bessel function; in the high frequency limit:

$$\text{gain} = 4\pi M \frac{1}{4\pi[2\omega D(M/R)^{1/2}]} \tag{6.116}$$

$$= \frac{2(MR)^{1/2}}{D}, \tag{6.117}$$

in agreement with the geometrical optics estimate. Notice that the on-axis behavior just considered is due to the long range behavior of Newtonian gravity.

The Bessel function behavior near the axis survives into the $R \to \infty$ behavior in $d\sigma/d\Omega$.

The divergence of the cross section for forward or backward direction gives rise to the scattering phenomenon called a *glory*. We now return to our concentration on the cross section and we continue by considering backward glories, where the deflection is close to π. Similar effects can be found also in the forward direction, caused by forward glories. Near the glory angle, rays with different impact parameters B exit into the same solid angle. Since the glory consists of those rays deflected through $\sim \pi$, a typical ray with impact parameter slightly greater than that appropriate for glory scattering will deflect slightly less than π. A second ray, incident on the opposite side of the axis with impact parameter slightly less than the glory angle, will deflect slightly more than π; it can be arranged that this deflection makes the second ray parallel to the first. Much of the subsequent parts of this chapter appeared in essentially this form in a paper by Matzner *et al.* (1985).

Although the glory phenomenon has long been well known in optical scattering (responsible for the bright halo one occasionally sees around the shadow of one's airplane) the definitive treatment for general potential scattering has been the series of articles by Ford *et al.* (1959) and by Ford & Wheeler (1959a,b). These papers are concerned with the semiclassical approximation to quantum scattering, but because of the close analogy with the scattering of massless waves in black hole gravitational fields, we can take them over completely. Mashoon (1974) considered electromagnetic glories in black hole spacetimes.

The general idea of the Ford–Wheeler approach is to examine the behavior of the classical function $\theta(l)$ at characteristic values and consider ensuing interference between l-modes of nearly the same value of θ. (Here l is the angular momentum per unit mass. We have for massless waves $l=\omega B$ where ω is the frequency of the waves and B is the impact parameter.) Since the classical cross section is $\sigma_{cl}=B\,\mathrm{d}B/(\sin\theta\,\mathrm{d}\theta)$, singularities can arise in the classical cross section from the inverse of $\sin\theta$. If $\theta(l)=\pi$ for some value of l, then there is a classical backward divergence. (For concreteness, we only consider backward glories.) The semiclassical analysis leads instead to a nondivergent glory cross section

$$\frac{\mathrm{d}\sigma_g}{\mathrm{d}\Omega}\propto J_0{}^2(l_g\sin\theta) \tag{6.118}$$

where J_0 is the Bessel function and l_g is the value of l at which θ passes through π.

Recently Nelson & DeWitt-Morette (1984) have given a much more compact and elegant derivation of the glory phenomenon, based on a JWKB version of the path integral for the full wave function. This extended JWKB formulation avoids the need for a separation into angular modes and their result is applicable to arbitrary, axially symmetric scattering problems. We will present a very heuristic version of the Nelson–DeWitt-Morette derivation, and then apply the results to scattering by black holes. We find there are striking differences between the scalar and the polarized cases; in particular for the polarized case there is in the JWKB limit a strong suppression of the exactly backward cross section, while the scalar case has a maximum in the backward direction.

In the classical case, if only the first ray were present, we would have (for an axisymmetric scatterer) (Goldstein, 1950)

$$\frac{\mathrm{d}\sigma}{\mathrm{d}\Omega}=\left|\frac{B\,\mathrm{d}B}{\sin\theta\,\mathrm{d}\theta}\right|. \tag{6.119}$$

The second ray contributes similarly to the cross section. Further, since $\theta=\pi$ is a regular point of the function $\mathrm{d}B/\mathrm{d}\theta$, we expect $\mathrm{d}B/\mathrm{d}\theta$ to be nearly the same for these two rays. Hence

$$\left.\frac{\mathrm{d}\sigma}{\mathrm{d}\Omega}\right|_{\substack{\text{classical}\\ \text{glory}}}=2\left|\frac{B\,\mathrm{d}B}{\sin\theta\,\mathrm{d}\theta}\right| \tag{6.120}$$

near the glory. The 2 indicates that two different classical paths contribute to the glory cross section; the $(\sin\theta)^{-1}$ indicates the divergence expected classically at the backward direction. When we admit wave interference between wave packets propagating along these two paths, the backward divergence is softened, and the characteristic interference pattern of a glory is, as we shall see, produced.

The expression for the glory scattering, derived by Nelson & DeWitt-Morette (1984) uses the full machinery of path integration techniques. In this particular case, the interference effects arise at a sufficiently low order that fairly unsophisticated argument leads to the correct formulation.

We will first consider the scattering of scalar quantum mechanical particles in an axisymmetric short-range force field confined near the origin of coordinates. We consider time independent systems only. The incident wave is a plane wave, initially travelling along the z axis toward smaller values of z:

$$\Psi_{\text{initial}} = \mathrm{e}^{-\mathrm{i}kz}\mathrm{e}^{-\mathrm{i}\omega t}, \tag{6.121}$$

where $k\hat{z} = \mathbf{k}_{\text{initial}}$ is the initial wave number, and ω is the (conserved) frequency of the wave.

In the WKB approximation, the final scattered wave is

$$\Psi_{\text{final}} \cong g(x)\mathrm{e}^{\mathrm{i}S}\mathrm{e}^{-\mathrm{i}\omega t}, \tag{6.122}$$

where S is the classical *action* (i.e., $\int p\mathrm{d}x$) along the ray, and g is an amplitude factor written explicitly below. The backward (glory) deflection is by assumption a possible classical path.

Away from the scatterer, we have simple free motion, so on this outgoing glory path $S = S_0 + k_{zg}z$ with k_{zg} constant. Here $S_0 + k_{zg}z_0$ is the contribution to the action accumulated up to some point z_0 on the outgoing orbit which is far from the scatterer.

The amplitude factor g can be shown to equal the ratio of infinitesimal volume elements defined by the rays at initial and final times. Such an infinitesimal volume element is made up by distinguishing incoming particles on rays spanning $\mathrm{d}\phi$, a small interval of ϕ; $\mathrm{d}B$, a small interval of impact parameter B; and $v_{\text{in}}\mathrm{d}t$, a small interval of length along the orbit. On emergence, they will be specified by some cross sectional area $\mathrm{d}A$ times a small interval of length, $v_{\text{out}}\mathrm{d}t$ along the outgoing orbits. Because we consider time independent fields, the time to traverse the orbit is exactly the same for each particle separated in the $v\mathrm{d}t$ direction along a particular path. Because of this, and because energy is conserved in time independent situations, the outgoing magnitude $|v_{\text{out}}\,\mathrm{d}t|$ is the same as the ingoing magnitude $|v_{\text{in}}\mathrm{d}t|$. Hence the amplitude factor g is also equal to the ratio of a cross section of initial rays, to the cross section they map into finally. The appearance of this factor is responsible, in non-glory situations, for yielding the classical cross section answer in the JWKB problem. Here, such an

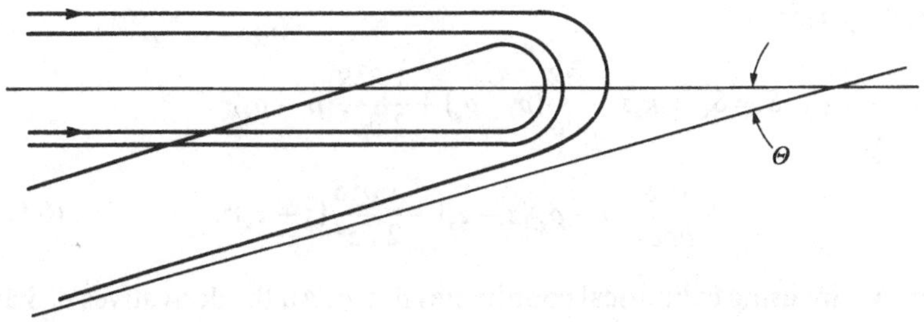

Fig. 6.4. When there is backward glory scattering, the deflection angle goes smoothly through (an odd multiple of) π while B passes through a finite range. Two rays with different B may then exit near the same backward direction θ.

answer would be infinite; furthermore we expect interference terms between different classical trajectories to modify the JWKB result.

We want to extract a cross section from this JWKB wave, at or near the glory angle. We need to find what amount of spherical wave is contained in the wave function, taking into account possible interference between rays. It is this consideration of interference between rays which makes this an *extended* JWKB calculation.

We proceed in a way which parallels the DeWitt-Morette–Nelson derivation. First of all, reference to fig. 6.4 shows that at large distance from the scatterer, and close to the glory angle, a bundle of rays incident with area

$$|B\,\mathrm{d}\phi\,\mathrm{d}B| \tag{6.123}$$

maps into an outgoing bundle with area

$$|B\mathrm{d}\phi \cdot R\mathrm{d}\,\theta|. \tag{6.124}$$

Hence

$$g \cong |\mathrm{d}B/R\,\mathrm{d}\theta|. \tag{6.125}$$

Still following the DeWitt-Morette–Nelson scheme, we now perform a three-dimensional Fourier transform on the outgoing wave. We are interested in directions near the glory angle, so we evaluate

$$\mathscr{K}(\mathbf{k}, \mathbf{k}_{\text{initial}}) = \mathrm{e}^{-\mathrm{i}\omega t} \int \mathrm{d}^3x \exp(-\mathrm{i}k_l x^l) g(x) \mathrm{e}^{\mathrm{i}S} \tag{6.126}$$

for $(k_l) = \mathbf{k}$ near the $+\hat{z}$ direction. We approximate the result in the following way. First of all, S can be expanded to quadratic order away from the glory ray:

$$S = S_0 + k_z z + \frac{\partial S}{\partial \rho}(\rho - \rho_g) + \frac{1}{2}\frac{\partial^2 S}{\partial \rho^2}(\rho - \rho_g)^2$$
$$+ \frac{\partial^2 S}{\partial \rho \partial z}(\rho - \rho_g)(z - z_g) + \frac{1}{2}\frac{\partial^2 S}{\partial z^2}(z - z_g)^2. \quad (6.127)$$

Here we are using cylindrical coordinates ρ, z, ϕ. All the derivatives of S are to be evaluated on the glory. The value of ρ_g equals B, the glory impact parameter, and will be so denoted. S is (up to a factor) the Hamilton–Jacobi principal function. We can use this fact from classical mechanics to deduce some of the properties of S, based on the standard result $p_i = \partial S/\partial x^i$. Because the glory trajectory is a classical path and ρ is constant asymptotically on the glory, $p_\rho = \partial S/\partial \rho$ vanishes asymptotically on the glory, as does $\partial^2 S/\partial \rho \partial z$. Because of the free propagation (with constant momentum) in the final ray, $\partial p^z/\partial z = \partial^2 S/\partial z^2 = 0$ on the glory also.

Now consider

$$\partial S/\partial \rho = k_\rho, \quad (6.128)$$

the ρ-component of the momentum, which, as we have said, vanishes on the glory, but not necessarily off of it.

For a ray travelling at a slight angle to the glory, we have asymptotically: $k_\rho = |k_{zg}|\rho/R \cong |k_{zg}|\rho/z$. So

$$\left.\frac{\partial^2 S}{\partial \rho^2}\right|_{\text{glory}} = \left.\frac{\partial}{\partial \rho}k_\rho\right|_{\rho=0} = \frac{|k_{zg}|}{R}. \quad (6.129)$$

Now let us compute (6.126) for some $\mathbf{k}$ that makes an arbitrary small (or zero) angle θ with the glory direction, but which has vanishing y momentum. The quantity $-ik_l x^l$ is then

$$-ik_l x^l = -ik_z z - ik_x x = -ik_z z - ik_x \rho \sin\phi. \quad (6.130)$$

The integral in (6.126) is over all values of $\rho \in (0, \infty)$, $z \in (-\infty, \infty)$, $\phi \in [0, 2\pi)$ and will involve an integration by the method of stationary phase. We perform this integration when $|k_x/k_z| \ll 1$ using the approximate expansion for S as obtained above

$$\mathcal{K}(\mathbf{k}, \mathbf{k}_{\text{initial}}) \cong e^{-i\omega t}\int e^{-ikz} g^{1/2} e^{iS}\left(\int \exp(-i\rho k_x \sin\phi)\, d\phi\right)\rho\, d\rho\, dz$$
$$= e^{-i\omega t}\int e^{-ikz} g e^{iS} 2\pi J_0(k_x \rho)\rho\, d\rho\, dz, \quad (6.131)$$

where the Bessel function arises from a standard integral formulation. Using the expansion (6.127)–(6.129) for S:

$$\mathscr{K}(\mathbf{k}, \mathbf{k}_{\text{initial}}) = \mathrm{e}^{-\mathrm{i}\omega t} 2\pi \mathrm{e}^{\mathrm{i}S_0} \int \exp\left[\mathrm{i}(k_{zg} - k_z)z\right] \mathrm{d}z$$

$$\times \int \exp\left(\mathrm{i}\frac{k_{zg}}{2R}(\rho - B)^2\right) g^{1/2} J_0(k_x \rho)\rho \, \mathrm{d}\rho. \qquad (6.132)$$

By assumption k_{zg} is large, and the ρ integral can be evaluated by the method of stationary phase:

$$\int_0^\infty \exp\left(\mathrm{i}\frac{k_{zg}}{2R}(\rho - B)^2\right) g^{1/2} J_0(k_x \rho)\rho \, \mathrm{d}\rho$$

$$\cong B(g^{1/2}(\rho = B)) J_0(k_x B) \int_{-\infty}^{\infty} \exp\left(\mathrm{i}\frac{k_{zg}}{2R}(\rho - B)^2\right) \mathrm{d}\rho$$

$$= B\left|\frac{\mathrm{d}B}{R\,\mathrm{d}\theta}\right|^{1/2} J_0(k_z B \sin\theta)\left(\frac{R}{k_{zg}}\right)^{1/2} (2\pi \mathrm{i})^{1/2}. \qquad (6.133)$$

The z-integration gives simply $2\pi\delta(k_{zg} - k_z)$, so to the order of our approximation

$$\mathscr{K}(\mathbf{k}, \mathbf{k}_{\text{initial}}) = (2\pi)^2 \delta(k_{zg} - k_z)\left(\frac{2\pi \mathrm{i}}{k_{zg}}\right)^{1/2} B\left|\frac{\mathrm{d}B}{\mathrm{d}\theta}\right|^{1/2} J_0(k_z B \sin\theta) \mathrm{e}^{\mathrm{i}S_0} \mathrm{e}^{-\mathrm{i}\omega t}. \qquad (6.134)$$

It will be noted that we have assumed that it is sufficiently accurate to consider only the asymptotic outgoing region in performing the Fourier transform that leads to (6.134). This is in some sense plausible, since in this way we include a large volume over which contributions to the chosen outgoing plane wave are added up and furthermore, we would expect a wave outgoing in the direction of interest to have its principal weight from this particular asymptotic region. The precise justification has been given (for nonrelativistic quantum mechanical scattering) by DeWitt-Morette & Zhang (1983) especially their section B(i).

The quantity $\mathscr{K}(\mathbf{k}, \mathbf{k}_{\text{initial}})$ calculated here has many names and many interpretations. It is clearly the $\mathbf{k}$ Fourier component of the scattered wave. (Because we assume an initial plane wave, there is no overlap except exactly in the forward direction with the incident wave.) Our notation for this Fourier component is chosen consistently with the notation for one of these interpretations:

$\mathscr{K}(\mathbf{k}, t; \mathbf{k}_{\text{initial}}, t_{\text{initial}})$ is the nonrelativistic quantum-mechanical probability amplitude that the wave with wavenumber $\mathbf{k}_{\text{initial}}$ at the early time t_{initial} will be found with wavenumber $\mathbf{k}$ at t (some late time). There is a standard prescription for extracting cross sections. The standard non-relativistic form is (slightly paraphrased from Nelson & DeWitt-Morette (1984)):

> $d\sigma/d\Omega$ is the transition probability into the direction Ω per unit time per unit incident flux. Hence we divide $|\mathscr{K}(\mathbf{k}, \mathbf{k}_{\text{initial}})|^2$ by $|t - t_{\text{initial}}|$ and by $|k|$, and sum over all states having the direction $\hat{\mathbf{k}}$. Since the sum over all $\mathbf{k}$ states is the integral over $\mathbf{k}$ with respect to the measure $d^3k/(2\pi)^3$ we obtain
>
> $$\frac{d\sigma}{d\Omega} = \frac{1}{|t - t_{\text{initial}}|}\frac{1}{|k|}\int_0^\infty \frac{k^2}{(2\pi)^3}|\mathscr{K}(\mathbf{k}, \mathbf{k}_{\text{initial}})|^2\, dk. \qquad (6.135)$$

The term $\delta(k_{zg} - k_z)$ may then be interpreted in terms of the δ-function conserving energy, and the square of this δ-function may be interpreted in a standard way as $|t - t_{\text{initial}}|^2$ (see Schiff, 1968). One obtains

$$\left.\frac{d\sigma}{d\Omega}\right|_{\text{near glory}} = 2\pi k_{zg} B^2 \left|\frac{dB}{d\theta}\right| J_0{}^2(k_{zg} B \sin\theta). \qquad (6.136)$$

Here we justify this procedure and result by a more directly wave-mechanical derivation. The calculation to (6.134) found the plane-wave Fourier component in a certain outgoing direction. We require the outgoing spherical wave component of the radiation. The plane wave can be represented in terms of spherical waves as (Messiah, 1958):

$$e^{i\mathbf{q}\cdot\mathbf{r}} \underset{r\to\infty}{\sim} \frac{2\pi}{iqr}[\delta(\Omega_r - \Omega_q)e^{iqr} - \delta(\Omega_r + \Omega_q)e^{-iqr}] + O(r^{-2}) \qquad (6.137)$$

i.e. the plane wave is dominantly an outgoing spherical wave near the direction that it is outgoing, and dominantly an ingoing spherical wave near the direction it is ingoing. The δ-functions here are defined as

$$\int \delta(\Omega' - \Omega)\varphi(\Omega')\, d\Omega' = \varphi(\Omega). \qquad (6.138)$$

Thus our Fourier decomposition into plane waves (6.134) can just as well be considered (asymptotically) a decomposition into spherical waves:

$$\Psi(r, t) \underset{r\to\infty}{\sim} \int \frac{d^3k}{(2\pi)^3}\mathscr{K}(\mathbf{k}, \mathbf{k}_{\text{initial}})\frac{2\pi}{ikr}[\delta(\Omega_r - \Omega_k)e^{ikr} - \delta(\Omega_r + \Omega_k)e^{-ikr}]$$

$$= e^{-ikr}\int d\Omega_k\, dk \frac{k^2}{ikr}\left(\frac{2\pi i}{k}\right)^{1/2} B\left|\frac{dB}{d\theta}\right|^{1/2} \qquad (6.139)$$

$$\times J_0(kB\sin\theta)[\delta(\Omega_r - \Omega_k)\delta(k_{zg} - k_z)e^{ikr} - \delta(\Omega_r + \Omega_k)\delta(k_{zg} - k_z)e^{-ikr}]e^{iS_0}e^{-i\omega t} \qquad (6.140)$$

where θ is the angle associated with the wave vector $\mathbf{k}$. Thus

$$\Psi \underset{r\to\infty}{\sim} \frac{1}{r} e^{-i\omega t} e^{ikr} \left(\frac{2\pi k}{i}\right)^{1/2} B \left|\frac{dB}{d\theta}\right|^{1/2} J_0(\omega B \sin\theta) e^{iS_0}$$
$$+ \text{contribution} \propto e^{-i\omega t} e^{-ikr}, \tag{6.141}$$

where now θ is the direction to the observation point. (We cannot explicitly evaluate the terms $\propto e^{-ikr}/r$ because they involve the opposite side of the sphere from our observation point. However, these terms are not needed to evaluate the cross section). The quantum-mechanical flux for an incident plane wave $e^{-ikz} e^{-i\omega t}$ is

$$|j^z| = |k|. \tag{6.142}$$

For the outgoing wave in (6.141) one has for the radial flux

$$|j| = 2\pi r^{-2} k^2 B^2 \left|\frac{dB}{d\theta}\right| J_0{}^2(kB \sin\theta). \tag{6.143}$$

The cross section is the ratio of these fluxes times r^2

$$\left.\frac{d\sigma}{d\Omega}\right|_{\text{near glory}} = 2\pi k B^2 \left|\frac{dB}{d\theta}\right| J_0{}^2(kB \sin\theta), \tag{6.144}$$

in agreement with (6.136).

6.2.3.1 *The attractive Coulomb problem*

Our goal is to verify formulae like (6.144) for the scattering of waves from a spherical black hole. We are guided by an analysis of scattering in the attractive Coulomb potential, and in a modified Coulomb potential which has an additional short range attractive force.

In one of the first quantum mechanical studies of the scattering by a Coulomb field Gordon (1928) includes a treatment in terms of the JWKB approximation, and, in particular, integrates up the phase along classical rays. We present an extract of the relevant part of his work.

In (nonrelativistic) classical mechanics, for an attractive Coulomb field, there is a conserved Hamiltonian

$$E = H = p^2/2m - q^2/r \tag{6.145}$$

(the scattered point mass has charge q, the fixed scattering center has charge $-q$.) There is of course the direct analogy to Newtonian gravity.

The particle moves in the potential at a varying speed:

$$v = v_0(1 + q^2/Er)^{1/2}, \tag{6.146}$$

where $v_0{}^2 = 2E/m$ is the velocity at infinity and it is useful to introduce $A = |A| = q^2/(2E) > 0$ for an attractive potential.

If $0 < B$ is the impact parameter for the incident particle, then the orbit is described by

$$r = \frac{B^2}{B\sin\theta + |A|(1+\cos\theta)}; \tag{6.147}$$

when $\theta = \theta_{in} = -\pi$, $r = \infty$. There is a point of closest approach at

$$0 < B/|A| = \tan\theta_{cl}. \tag{6.148}$$

The scattered particle again reaches infinity when the denominator vanishes

$$0 > -|A|/B = \tan(\tfrac{1}{2}\theta_{out}). \tag{6.149}$$

The action is

$$S = m\int v\,\mathrm{d}s, \quad \mathrm{d}s = \text{element of arclength}. \tag{6.150}$$

As can be seen from fig. 6.5, $\mathrm{d}s = |\mathrm{d}y/\sin\tau|$, where τ is the angle that the ray makes with the z-axis. By a series of algebraic manipulations we can evaluate the integral. The equation of the hyperbola is

$$\mathrm{A}(r+z) = B(B-y), \tag{6.151}$$

which gives

$$|\sin\tau| = \left|\frac{v_0}{v}\frac{B-y}{r}\right|, \tag{6.152}$$

so, using (6.151) again

$$S = -m\int\frac{v_0(r+2|A|)}{(B-y)}\mathrm{d}y. \tag{6.153}$$

Further, substituting $z^2 = r^2 - y^2$ in (6.151) gives

$$r = \frac{A^2+B^2}{2|A|B}(B-y) - |A| + \frac{|A|B}{2(B-y)}, \tag{6.154}$$

so the integral for the phase can be done explicitly:

$$S = mv_0\left(\frac{A^2+B^2}{2|A|B}(B-y) + |A|\ln\left|\frac{y-B}{B}\right| + \frac{|A|B}{2(y-B)}\right). \tag{6.155}$$

In order to obtain a definite result like (6.155) for the action S, one must fix the lower limit in the integral. (The upper limit is the field point, (y, z).) We

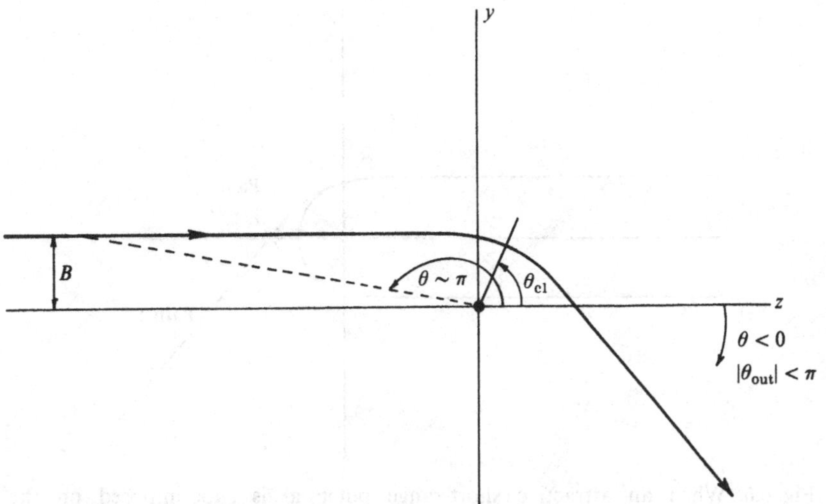

Fig. 6.5. For attractive Coulomb scattering with incident impact parameter B, the angle at which closest approach occurs, θ_{cl} always satisfies $\theta_{cl} \geqslant 0$; the exit angle satisfies $-\pi \leqslant \theta_{cl} < 0$; equality in each case occurs when $B = 0$ exactly.

have followed Gordon (1928) and taken the constant so that the action is $-mv_0B^2/2|A|$ at the point where the ray crosses the z-axis. The choice adds, obviously, a B-dependent constant to the phase, but has only a small effect on the phase at large distances in a particular direction.

Notice that in most directions, according to (6.154),

$$r \sim \frac{A^2 + B^2}{2|A|B}(B - y). \tag{6.156}$$

The exception occurs when $|B - y|$ in the outgoing ray is small. This special case will merit some consideration below, but for most directions, the scattered phase is

$$S \sim mv_0(r + |A| \ln r + \mathrm{O}(\mathrm{const})). \tag{6.157}$$

This logarithmic addition to the phase is characteristic of Coulomb scattering, and is, as we have seen, the analog of the familiar behavior of waves in the vicinity of a black hole.

6.2.3.2 Attractive Coulomb plus short-range attractive potential

The analysis in section 6.2.3.1 gives the complete behavior of the phase near infinity for the Coulomb case. The Coulomb potential is not a full model for

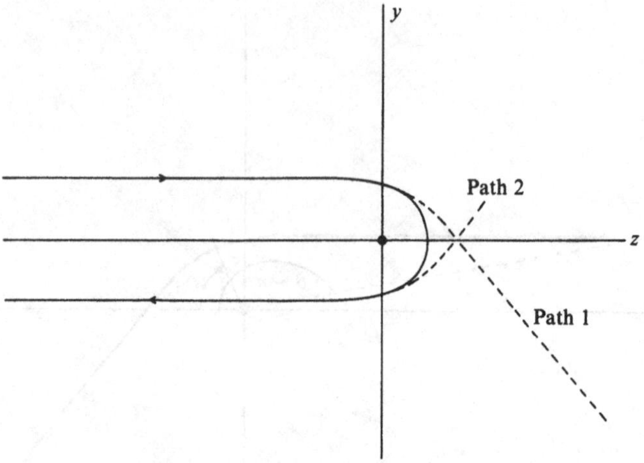

Fig. 6.6. When an attractive short-range potential is superimposed on the Coulomb scattering case, glories will occur. Any ray that enters the potential and approaches closely enough to feel the additional short range force will undergo an extra deflection and so may emerge in the exactly backward direction even though its impact parameter is nonzero. In fig. 6.6, the incident ray, path 1 would emerge on the extended dotted path, if the motion were in a pure Coulomb field; similarly, the emerging ray would have come from the indicated dotted direction path 2, if the field were really pure Coulomb. The effect of the short range force is thus to join together to Coulomb-orbit segments, one ingoing, one outgoing.

the scattering for black holes, because it has no glories. Only rays that enter with (in the limit) vanishing impact parameter exit exactly in the backward direction, $dB/d\theta \to 0$ in this limit, and the backward cross section is finite.

By considering an additional short-range potential, we can, however, have glories, see fig. 6.6. The additional short-range force increases the deflection angle, so that backward scattering occurs for finite (nonzero) impact parameter. Notice that the exceptional case alluded to in connection with (6.156) above, $(B - y)$ small on the outgoing ray, apparently occurs in the backward scattering direction, and in that direction the formula (6.157) is suspect. Here we have backward scattering. But notice that in (6.151) *et seq.*, B is the (asymptotic) distance from the axis, as measured for the *incoming ray.* With the added short-range force, the outgoing ray is asymptotically a segment of an outgoing Coulomb scattered ray, but one that did not in the context have small impact parameter, and one that does not exit near the direction it (would have, if extended backwards in the Coulomb solution) started from. Hence, we can safely say that the dominant part of the action for these glory scattering long-range forces is

$$S \sim mv_0[r + |A| \ln(r/r_0)]. \tag{6.158}$$

6.2.3.3 Glory scattering for long-range forces

To compute the glory scattering for these long-range forces, we notice that the phase in the JWKB approximation to quantum mechanics is $iS/\hbar$; and since we will take $\hbar = 1$, we have $mv_0 = k$, the wave number (at infinity). Hence the asymptotic phase is $\sim kr^*$, where

$$r^* \equiv [r + |A| \ln (r/r_0)], \tag{6.159}$$

and the normalizing (integration constant) r_0 is chosen as convenient.

Although in section 6.2.3.1 specific meaning was given to y and z (referring to the coordinates for the exact Coulomb orbits) here we use them as ordinary coordinates again, for our long-range glory scattered rays. Then, on the glory ray (asymptotically):

$$\Psi = e^{-i\omega t} g^{1/2} e^{i(kz^* + S_0)}, \tag{6.160}$$

since the wave is travelling in the z^* direction, so $z^* = r^*$ in this case. Here g is the same as in (6.125). Our aim in this section is to show that the presence of the logarithmic shift in (6.160) leaves the calculation of the cross section unchanged; the corrections in the calculation appear at an order that is irrelevant for objects, like the cross section, that are evaluated at spatial infinity.

First of all we have, far from the scatterer,

$$\partial S/\partial z^* = k. \tag{6.161}$$

To evaluate the second derivative, we use the fact that we know the phase to one order beyond our definition of z^* (or r^*) in (6.159). In fact, the phase on the glory orbit goes

$$k(z + \ln z + \text{const} + O(1/z)) = kz^*(1 + O(\text{const}/z^*)).$$

So

$$\partial^2 S/\partial z^{*2} = O(\text{const}\, k/z^{*2}) \tag{6.162}$$

and will be neglected.

Also, if we now introduce

$$\rho^* = r^* \sin\theta \quad \text{to complement } z^* = r^* \cos\theta, \tag{6.163}$$

we have

$$k_\rho{}^* = \partial S/\partial \rho^* = 0, \tag{6.164}$$

because this is a classical path with asymptotically no ρ^* motion. We further proceed, as in section 6.2.3.

$$\frac{\partial}{\partial z^*}\frac{\partial S}{\partial \rho^*} = \frac{\partial}{\partial z^*} k_\rho{}^* = \frac{\partial}{\partial z^*} k \frac{\rho^*}{z^*} \propto k \frac{\rho^*}{z^{*2}}, \tag{6.165}$$

but

$$\frac{\partial^2 S}{\partial \rho^{*2}} = \frac{\partial}{\partial \rho^*} k_\rho{}^* = \frac{k}{z^*}, \tag{6.166}$$

which in analogy to the result in (6.129) is the dominant derivative. Because the difference between r and r^* is only logarithmic, the differences in derivatives yield *only* differences that vanish at least $O(r^{-1})$ compared to the dominant terms in each derivative. Hence the phase has the form given by (6.127)–(6.129), but with z^* and ρ^* replacing z and ρ.

We now proceed with an expansion in terms of a Fourier transform utilizing an expansion in 'distorted plane waves' $e^{i\mathbf{k}\cdot\mathbf{r}^*}$. We make the same assumption as in section 6.2.3 that we sample enough of the outgoing wave in the asymptotic region that this provides a sufficiently accurate transform.

Notice that in doing this the angles θ, ϕ are still well defined because of the spherical symmetry. We can, just as before, express each distorted plane wave (asymptotically) in terms of spherical modes like (6.137):

$$e^{i\mathbf{q}\cdot\mathbf{r}^*} \underset{r^*\to\infty}{\sim} \frac{1}{iqr}[e^{iqr^*}\delta(\Omega_q - \Omega_{r*}) - e^{-iqr^*}\delta(\Omega_q - \Omega_r)][1 + O(\ln r/r)]. \tag{6.167}$$

The corrections due to the logarithmic behavior decay away rapidly enough at infinity so that the cross section can be extracted, and we obtain exactly (6.136).

6.2.3.4 *The scattering of massless scalar waves by a Schwarzschild 'Singularity'* (Matzner, 1968)

Recall from chapter 1 that the propagation of massless scalar waves in the Schwarzschild field is described by the curved-space wave equation,

$$0 = \Box\Phi = \Phi^{;\alpha}{}_{;\alpha} \tag{6.168}$$

and in the Schwarzschild geometry the solution is separable into time, angular, and radial factors:

$$\Phi = \sum_{lm} \int d\omega e^{-i\omega t} R_{l\omega}(r) Y_l^m(\theta, \phi). \tag{6.169}$$

Recall also that the asymptotic radial wave function has the behavior $e^{i\omega r^*}/r$. Finally, we have shown in section 6.2.1 that in the JWKB limit, the normal to the phase fronts (i.e. the vector $k^\alpha = \Phi^{;\alpha}$) is geodesic, which is to say that the JWKB solution is the eikonal solution for the motion of massless particles in classical mechanics. We thus have complete analogy with the wave mechanical problem considered above in sections 6.2.3–6.2.3.3. We wrote the previous discussion in a way that referred only to the wave properties (not to their physically equivalent mechanical counterparts in quantum mechanics) so that this transition would be immediate. The only point to be checked is the definition of the flux in determining the cross section. For massless plane waves we use, instead of (6.142), the stress tensor, (5.5).

The incident flux T_{0z} is then proportional to ω^2, rather than to k, as in the quantum mechanical formula (6.142). The radially outgoing flux is similarly proportional to ω^2. The factors of ω^2 in this formula (and factors $|k|$ in the quantum formula) then cancel in any case.

We conclude then, that the glory scattering cross section for scalar massless waves is identically (6.136) where now $|k|$ means ω. All references in that equation are to quantities that can be measured (or specified) at infinity, so we can completely determine the glory.

For scattering from a spherical black hole, the function $\theta(B)$, for instance, has a simple analytical limiting form (Darwin, 1959, 1961; Misner, *et al.*, 1973) when $\theta \gtrsim \pi$:

$$B/M \sim 3\sqrt{3} + 3.48\mathrm{e}^{-\theta}. \tag{6.170}$$

The first backward glory $\theta = \pi$ occurs roughly at

$$B/M - 3\sqrt{3} = 3.48\mathrm{e}^{-\pi}, \tag{6.171}$$

$$\cong 0.15; \tag{6.172}$$

also, from (6.170)

$$\frac{\mathrm{d}B}{\mathrm{d}\theta} \cong -3.48Me^{-\theta} \tag{6.173}$$

$$\cong -0.15M \quad \text{when } \theta = \pi. \tag{6.174}$$

We thus have

$$\left.\frac{\mathrm{d}\sigma}{\mathrm{d}\Omega}\right|_{\text{glory}} \cong 2\pi\omega M(3\sqrt{3} + 3.48\mathrm{e}^{-\theta})^2 M^2(3.48\mathrm{e}^{-\theta}) \times J_0{}^2[\omega M(3\sqrt{3} + 3.48\mathrm{e}^{-\theta})\sin\theta]. \tag{6.175}$$

Since $B^2\,\mathrm{d}B/\mathrm{d}\theta$ appears in the glory cross section, we see that the contribution from the second backward glory is at least a factor $\mathrm{e}^{-2\pi} = 2 \times 10^{-3}$ smaller than the first, and each decreases by such a factor, so we consider only the first backward glory, $\theta = \pi$. Hence

$$\left.\frac{\mathrm{d}\sigma}{\mathrm{d}\Omega}\right|_{\text{glory}} \cong 2\pi\omega M M^2(28.59) J_0{}^2[\omega M(5.35)\sin\theta]. \tag{6.176}$$

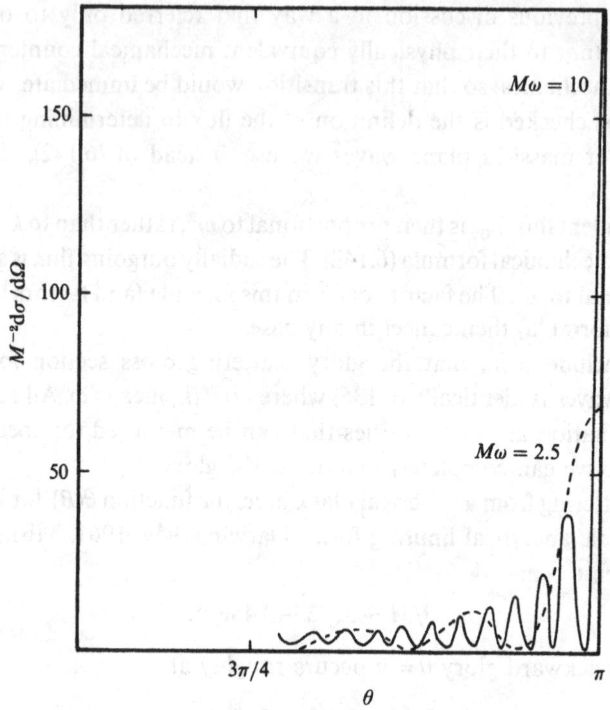

Fig. 6.7. When the JWKB cross-section form (6.136) is computed for the backward direction in black hole scattering of scalar waves, there results an oscillatory behavior peaked at the axis. Here we show the result for $M\omega = 2.5$ and for $M\omega = 10$. We expect that the JWKB results will be valid only for $|\theta - \pi| \ll 1$, and only for $M\omega \gg 1$. Both examples shown here violate those conditions, but the examples will be useful for comparison to numerical results in chapter 8.

The geometrical (high frequency) *capture* cross section of a black hole is $27\pi M^2$. This backward cross section is modified in (6.176) principally by the factor arising from $dB/d\theta$, and diverges as the radiation wavelength $\lambda \to 0$ ($\omega \to \infty$). As is well known, this kind of glory scattering reproduces the classical result, including the factor of two in (6.120), if we average over the oscillations of $J_0{}^2$. Fig. 6.7 shows the cross sections obtained near the backward direction, applying this formula in the two cases, (*a*) $M\omega = 2.5$; (*b*) $M\omega = 10$. The former parameter is sufficiently small that JWKB results should be suspect (as, in fact, is the $M\omega = 10$ case), but it will be interesting to compare it with some results on polarized scattering, cf below. We have shown the computed cross section for a range almost $\frac{1}{4}\pi$ from the backward direction. Our approximations in deriving (6.176) are certainly poor over so large a range, but these results will be qualitatively useful in our analysis in chapter 8.

6.2.3.5 *Polarized wave glory scattering by black holes*

In a general scattering problem the propagation of the polarization is determined by the specifics of how the spin interacts with the potential, and this can conceivably be very complicated. For the propagation of massless radiation in the high frequency limit we have already found (cf sections 6.2.1 and 6.2.2) the very simple JWKB rule: *polarization is parallel transported along the orbit.*

Consider the glory scattering of a massless quantum in the Schwarzschild geometry. The orbit lies in a plane, and is sketched in fig. 6.8(a). We can consider the parallel propagation of two different spacelike vectors orthogonal to the propagation vector: (A), perpendicular to the (two-dimensional) plane of the orbit; the other (B), lying in this plane. The first case, the vector perpendicular to the plane of the orbit, can be easily seen by symmetry, to maintain that orientation. A more interesting problem is prescribed by the in-plane vector.

To investigate this case, we note that we can introduce, in addition to the null propagation vector $\mathbf{k}$, a second null vector $\mathbf{l}$ (also with future-pointing time part). They may be relatively normalized by $\mathbf{l}\cdot\mathbf{k} = -2$ (say). Suppose we pick the orbital plane to be the $\phi = 0$ orbit. Then vectors $\mathbf{e}$ that lie in this plane have $\mathbf{e}\cdot\partial/\partial\phi = 0$. Parallel propagation preserves the scalar product. Consider a specific spacelike vector $\mathbf{e}$ which is transverse to the orbit, and lies in the orbital plane, and has unit length, all specified at the instant of peribarythron; see the vector labelled B^* in fig. 6.8(a). This vector satisfies

$$\mathbf{e}\cdot\mathbf{e} = 1, \tag{6.177}$$

$$\mathbf{e}\cdot\partial/\partial\phi = 0, \tag{6.178}$$

$$\mathbf{e}\cdot\mathbf{l} = 0, \tag{6.179}$$

$$\mathbf{e}\cdot\mathbf{k} = 0, \tag{6.180}$$

where

$$\mathbf{l}\cdot\mathbf{l} = 0, \tag{6.181}$$

$$\mathbf{k}\cdot\mathbf{k} = 0, \tag{6.182}$$

$$\mathbf{l}\cdot\mathbf{k} = -2. \tag{6.183}$$

All these relations are preserved under parallel transport; at spatial infinity, they give $\mathbf{e}_{\text{in}} = \partial/\partial y$; $\mathbf{e}_{\text{out}} = -\partial/\partial y$. (The transported vector must

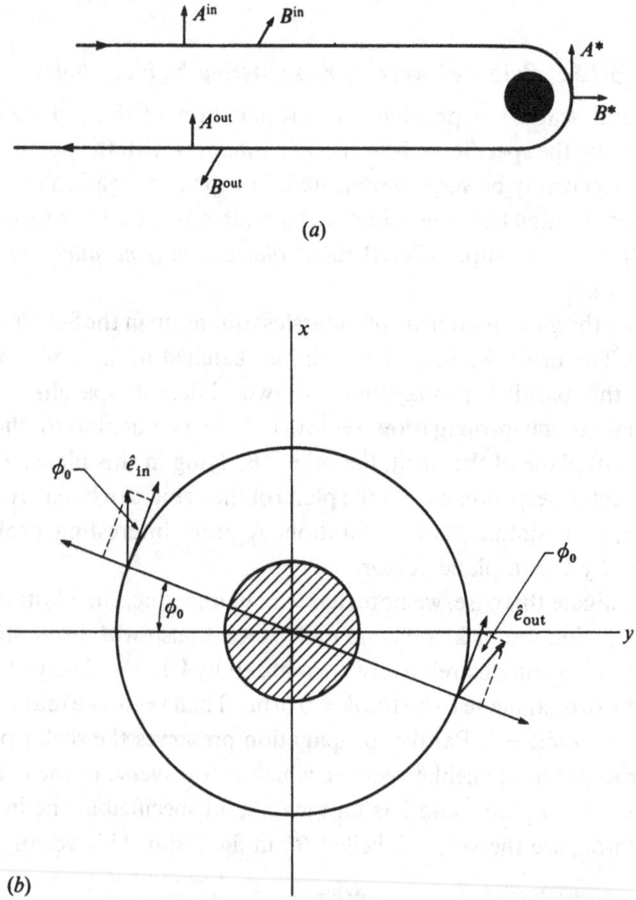

Fig. 6.8.(*a*) Linear polarization vectors are parallelly propagated along null particle orbits in the Schwarzschild geometry. For vectors like A, which are initially perpendicular to the orbital plane, the vector emerges into the asymptotic flat region parallel to its incident direction. Vectors like B which are initially in the plane of the orbit emerge also in the plane but pointing in the opposite sense. (*b*) When the plane of the orbit is tilted so that an initial polarization (on the left, parallel to the x-axis) makes an angle ϕ_0 to the plane of the orbit, then one finds that after parallel transport around the black hole, the polarization has undergone a net rotation (measured in the asymptotic flat region) of $2\phi_0$. This is relevant because in glory scattering one has a whole plane wave-front of orbits, all of whose polarizations initially are aligned, say with the x-axis. After orbiting, the typical ray has an associated rotated polarization, so the integrations carried out to obtain the backward glory contribution in fact yield vanishing contribution in the backward direction.

always point in the spacelike direction to the outside of the orbit, i.e. $e^r > 0$, because to 'cross' the orbit would require $\mathbf{e}\cdot\mathbf{e} = 0$, in violation of (6.177).)

Hence we have that for glory scattering, spacelike vectors parallely propagated, which lie initially in the plane of the orbit, emerge still in the plane of the orbit, but point in the opposite sense. Spacelike vectors parallelly propagated, which initially are perpendicular to the plane of the orbit, emerge still perpendicular, with the same sense.

In our problem we have to consider a congruence of orbits, incident at different places around the scatterer, all of which are deflected back into the glory. Suppose each parallely propagates a spacelike vector, and that in the incident congruence, these spacelike vectors are all mutually parallel (parallel to the x-axis, say). Fig. 6.8(*b*) shows the situation after the ray is deflected to the glory angle.

Because different rays orbit the black hole in different planes specified by an angle ϕ_0, the parallel transported spatial vectors do not all emerge mutually parallel, but in fact are rotated by the amount $2\phi_0$.

For a classical spin-s radiation field ψ (written in complex notation, i.e. $\psi \sim e^{i\omega t} e^{i\omega z}$), rotating by an angle α about its propagation direction gives a new value (Misner, *et al.*, 1973) for ψ:

$$\psi' = e^{is\alpha}\psi; \tag{6.184}$$

the effect of a rotation of the emerging field is to introduce the phase shift $s\alpha$, and we have just seen that parallel propagation leads to rotations of angle $\alpha = 2\phi$.

With this information we can continue the analysis of the nonzero glory scattering. We assume the incident wave has some uniform (flat space parallel) polarization and then we have

$$\psi_{\text{final}} = g^{1/2} e^{iS} e^{-i\omega t} e^{2i\phi s}, \tag{6.185}$$

where ϕ is the cylindrical coordinate polar angle to the point at which ψ_{final} is evaluated, and s is the spin of the particle.

We now proceed as in section 6.2.3 to perform the Fourier transform integration. Equation (6.131) is replaced by

$$\mathscr{K}_s(\mathbf{k}, \mathbf{k}_{\text{initial}}) \cong e^{-i\omega t} \int e^{-ikz^*} g^{1/2} e^{iS} \int_0^{2\pi} e^{2is\phi} \\ \times \exp(-ik_x \rho \sin\phi)\, d\phi\, \rho^*\, d\rho^*\, dz^*, \tag{6.186}$$

where the 'starred' variables arise as demonstrated in sections 6.2.3.1–

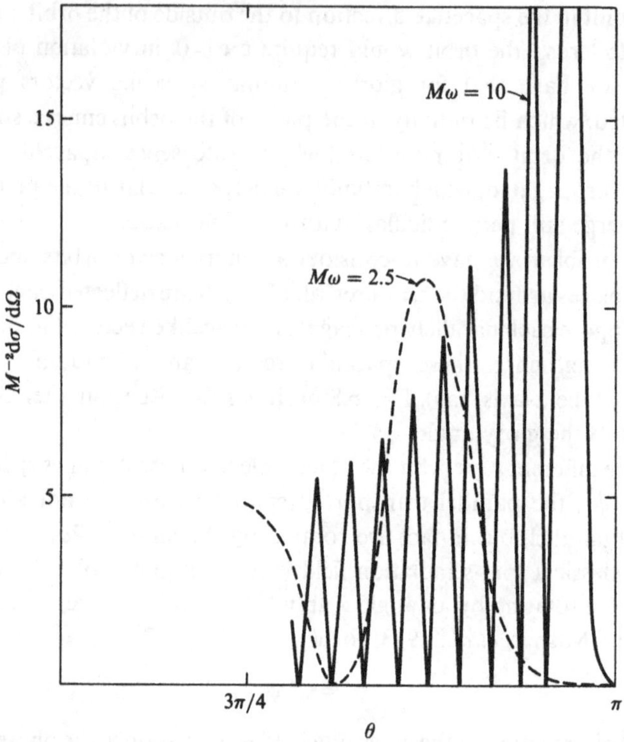

Fig. 6.9. The spin-2 JWKB glory approximation for scattering by a black hole given by (6.188) with (6.170)–(6.174) differs from the spin-0 case since its dominant angular dependence is $J_4^2(\omega B \sin\theta)$, while that of the spin-0 case is $J_0^2(\omega B \sin\theta)$. Thus the gravitational (and other nonzero spin) cases have vanishing cross section (in this JWKB limit) in the backward direction. Here we plot the $M\omega = 2.5$ and $M\omega = 10$ glory behaviors near the background direction.

6.2.3.4 due here to the long-range behavior of gravity. The ϕ integral gives a standard definition (Gradshteyn & Ryzhik, 1965)

$$\int_0^{2\pi} \exp\left[i(2s\phi - k_x\rho \sin\phi)\right] d\phi = (-)^{2s} 2\pi J_{2s}(k_x\rho). \qquad (6.187)$$

The integration is thus identical to that described in section 6.2.3.5 for scalar waves except that $(-1)^{2s} J_{2s}$ replaces every appearance of J_0. The definition of fluxes is different for non-zero spin cases from the scalar case. For $s = \frac{1}{2}$ and $s = 1$, one can use the appropriate stress tensor components, (5.26) and (5.11), which in analogy to (5.5) are quadratic in the fields. For gravity, we use the Isaacson formula, (5.15), which is also quadratic. Hence the final, spinning particle, glory cross section is the generalization of (6.136)

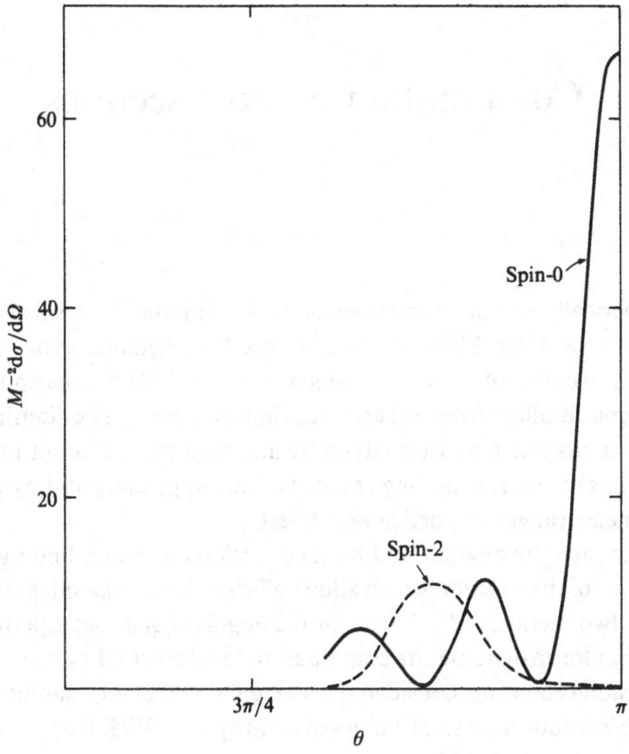

Fig. 6.10. Here we compare the spin-2 with the spin-0 black hole scattering cross sections, calculated when $M\omega = 2.5$ via the JWKB glory approximation. Although $M\omega = 2.5$ is a relatively low frequency to expect the validity of this approximation, we will show in chapter 8 that it in fact gives a good qualitative representation of the numerically computed cross section near the backward direction.

obtained by replacing $J_0 \to J_{2s}$

$$\left.\frac{\mathrm{d}\sigma}{\mathrm{d}\Omega}\right|_{\text{glory}} = 2\pi\omega B^2 \left|\frac{\mathrm{d}B}{\mathrm{d}\theta}\right| J_{2s}^2(\omega B \sin\theta). \tag{6.188}$$

Into this formula we insert the values of B, $\mathrm{d}B/\mathrm{d}\theta$ of (6.170)–(6.174).

Figure 6.9 shows the gravitational case approximation, (6.188), to black-hole scattering, for two values of ωM; $\omega M = 2.5$ and $\omega M = 10$. In both cases, the curves should be compared with the graphs for the scalar massless particle (fig. 6.7); in fig. 6.10 we plot the spin-0 and spin-2 $M\omega = 2.5$ cases for comparison.

7
Computation of cross sections

We numerically calculate cross sections for gravitational waves axially incident on a Kerr black hole. The resulting detailed cross sections display a wealth of structure which can be linked conceptually to phenomena familiar from other scattering problems. The limiting cross sections in chapter 6 assist analysis by allowing truncation of numerical calculations where the limiting results become applicable and by pointing to parameter ranges of particular interest.

We examine the angular and radial equations in detail, finding in each case a form of the solution which allows efficient numerical integration. We consider two methods of solution for the angular equation; a perturbation calculation for insight and its continuation for the actual numerical work. For the radial equation, considered in the remainder of the chapter, we use a slow stable solution for small values of r and apply a JWKB approximation to integrate rapidly to large r.

7.1 Angular equation

We first consider the perturbation of the angular equation about $a\omega = 0$, where as before, a is the specific angular momentum and ω is the scattered wave frequency.

One reason for considering the perturbation calculation is that it lays the foundation for the actual method used. In particular we follow a technique of Press & Teukolsky (1973) to expand the spin-weighted spheroidal harmonics conveniently in the spin weighted spherical harmonics in both cases. The perturbation calculation only obtains spheroidal values to first or second order in $a\omega$. The continuation method utilizes the same expansion of the spheroidal functions, but finds expressions for their derivatives with respect to $a\omega$. The numerical solution then integrates these derivatives.

Another reason for considering the perturbation calculation is that we will need to know the behavior of the spheroidal functions and eigenvalues as l becomes large in order to avoid truncating our mode sum at too small a value of l. We will find that the perturbation solution converges rapidly in $(a\omega)^2/l$ for $a\omega$ fixed and l large.

If the angular equation (2.27) is rewritten in the form

$$(\mathscr{L}_s + \mathscr{L}_1)_s S_l^m = -{}_s E_l^m {}_s S_l^m \tag{7.1}$$

with

$$\mathscr{L}_s(\theta) = \frac{1}{\sin\theta}\frac{\mathrm{d}}{\mathrm{d}\theta}\left(\sin\theta\frac{\mathrm{d}}{\mathrm{d}\theta}\right) - \left(\frac{m^2 + s^2 + 2ms\cos\theta}{\sin^2\theta}\right) \tag{7.2}$$

and

$$\mathscr{L}_1(a\omega, \theta) = (a\omega)^2\cos^2\theta - 2a\omega s\cos\theta \tag{7.3}$$

it is immediately amenable to a perturbation treatment for $a\omega$ small with a straightforward continuation to arbitrary values of $a\omega$ (Press & Teukolsky, 1973). We use the spin weighted spherical harmonics ${}_sY_l^m$ as a representation. They satisfy $\mathscr{L}_s Y_l = -E_l^{(0)} Y_l$ with $E_l^{(0)} = l(l+1)$, where the s, m indices are suppressed on E and Y but understood.

In view of the symmetries of the angular equation ((2.43)–(2.46)) we can consider only one of the spin values ± 2. For small $a\omega$ standard perturbation theory yields (cf Schiff, 1968)

$$E_l = l(l+1) - \langle Y_l | \mathscr{L}_1 | Y_l \rangle \tag{7.4}$$

and

$$\begin{aligned} S_l &= \sum_{l'} A_{ll'} Y_{l'} \\ &= Y_l - \sum_{l' \neq l} \frac{\langle Y_{l'} | \mathscr{L}_1 | Y_l \rangle}{l(l+1) - l'(l'+1)} Y_{l'}, \end{aligned} \tag{7.5}$$

where

$$\langle Y_l | W | Y_k \rangle \equiv \int_{\text{unit sphere}} \mathrm{d}\Omega \bar{Y}_l W Y_k.$$

The relation between the spin-weighted spherical harmonics Y_l and the rotation matrix elements of quantum mechanics (Campbell & Morgan, 1971) allows evaluation of the needed inner products $\langle Y_{l'} | \cos^2\theta | Y_l \rangle$ and $\langle Y_{l'} | \cos\theta | Y_l \rangle$ in terms of Clebsch–Gordon coefficients. The result is

$$\langle Y_{l'} | \cos^2\theta | Y_l \rangle = \frac{1}{3}\delta_{ll'} + \frac{2}{3}\left(\frac{2l+1}{2l'+1}\right)^{1/2} \langle l2m0 | l2l'm \rangle \langle l2-s0 | l2l'-s \rangle$$

and

$$\langle Y_{l'} | \cos\theta | Y_l \rangle = \left(\frac{2l+1}{2l'+1}\right)^{1/2} \langle l1m0 | l1l'm \rangle \langle l1-s0 | l1l'-s \rangle.$$

Hence to first order in $a\omega$ the eigenvalues are

$$E_l = l(l+1) - \frac{2a\omega s^2 m}{l(l+1)} \quad s \neq 0 \tag{7.6}$$

and the expansion coefficients $A_{ll'}$ in (7.5) are given by

$$A_{ll'} = \delta_{ll'} - \frac{\langle Y_{l'} | \mathscr{L}_1 | Y_l \rangle}{l(l+1) - l'(l'+1)}(1 - \delta_{ll'}). \tag{7.7}$$

For the case in which $a\omega$ is not small, we view $\mathscr{L} = \mathscr{L}_s + \mathscr{L}_1$ as an embedding of the operator $\mathscr{L}_s(\theta)$ in a one-parameter family of operators $\mathscr{L}(a\omega, \theta)$, with boundary conditions $\mathscr{L}(a\omega = 0, \theta) = \mathscr{L}_s$ and $\mathscr{L}(a\omega, \theta) = \mathscr{L}_s + \mathscr{L}_1 +$ '$O(a\omega)^2$' (cf Wasserstrom, 1972). The derivative with respect to $a\omega$ of (7.1), denoted by a dot, is

$$\dot{\mathscr{L}} S_l + \mathscr{L} \dot{S}_l = -\dot{E}_l S_l - E_l \dot{S}_l.$$

Taking the inner product of the above with $\langle S_{l'}|$ and using the self adjointness of $\mathscr{L}$, the orthonormality of S_l and Y_l, and the fact that $\dot{\mathscr{L}} = \dot{\mathscr{L}}_1$ one has

$$\langle S_{l'} | \dot{\mathscr{L}}_1 | S_l \rangle = (E_{l'} - E_l) \langle S_{l'} | \dot{S}_l \rangle - \dot{E}_l \delta_{ll'}. \tag{7.8}$$

For $l = l'$ one then has the continuation equation for E_l

$$\dot{E}_l = -\sum_{\alpha\beta} A_{l\alpha} A_{l\beta} \langle Y_\alpha | \dot{\mathscr{L}}_1 | Y_\beta \rangle. \tag{7.9}$$

The off-diagonal part of (7.8) provides the continuation equations for the coefficients $A_{l\gamma}$. If (7.8) is written in the form

$$\langle S_{l'} | \dot{S}_l \rangle = \langle S_{l'} | \dot{\mathscr{L}}_1 | S_l \rangle / (E_{l'} - E_l)$$

it is also clear that

$$\dot{S}_l = \sum_{l' \neq l} \frac{\langle S_{l'} | \dot{\mathscr{L}}_1 | S_l \rangle}{E_{l'} - E_l} S_{l'} \tag{7.10}$$

and therefore

$$A_{l\gamma} = \sum_{\substack{l' \neq l \\ \alpha, \beta}} A_{l'\alpha} A_{l\beta} \frac{\langle Y_\alpha | \dot{\mathscr{L}}_1 | Y_\beta \rangle}{E_{l'} - E_l} A_{l'\gamma}. \tag{7.11}$$

Note that the absence of a term proportional to S_l is consistent with the normalization condition $\langle S_l | S_l \rangle = 1$ implying $\langle S_l | \dot{S}_l \rangle = 0$. Once the α, β coefficients are obtained, the normalization condition

$$\sum_\alpha A_{\rho\alpha} A_{\tau\alpha} = \delta_{\rho\tau} \tag{7.12}$$

provides a further check on the accuracy of the integration.

Thus given the initial values $A_{ll'}(a\omega=0)=\delta_{ll'}$ and $E_l(a\omega=0)=l(l+1)$ from the known properties of the spin-weighted spherical harmonics, numerical integration of (7.9) and (7.11) provides a continuation to the spheroidal case for arbitrary $a\omega \neq 0$.

Equations (7.9) and (7.11) were in fact integrated in the numerical work by a fourth-order Runge–Kutta method. The integrated values are stored at intervals of $\Delta(a\omega)=0.2$ and a seventh-order polynomial fit generated for interpolation between the integrated values as needed (Press & Teukolsky, 1973; Handler, 1979; Handler & Matzner, 1980).

Note that the continuation above is a straightforward 'brute force' approach. For the range of values of $a\omega$ and l considered here this method is adequate.

To extend the calculation to larger values of $a\omega$ (requiring summing to larger values of l) or to consider off-axis scattering (requiring a sum over m for $-l \leqslant m \leqslant l$ for each value of l) would require an improved method. Intuitively we expect some form of JWKB approximation would apply. For larger values of l the eigenvalue in the angular equations (7.2)–(7.3) dominates over a large range of angle. In addition, the functions themselves oscillate rapidly in θ. This is the kind of situation in which a JWKB solution is applicable. One expects to match a JWKB solution for the midrange angles to the solutions valid at the endpoints $\theta=0$ and π.

7.2 Radial equation

We begin by applying the method of Press & Teukolsky (1973) to transform the radial equation (2.27) to a form in which both the ingoing and outgoing solutions possess the same r dependence asymptotically. This method has the disadvantage that the solutions oscillate rapidly in r, hence, to integrate the equation to large r we must use a small step size over a large range of r. Further, the equation in this form is unfortunately ill suited to JWKB approximation. The reason is shown to be the imaginary part of the $1/r$ term in the effective potential, which cannot be immediately removed without a reappearance of four powers of r difference in the resultant solutions. Chandrasekhar's (1975b, 1978a–c) transformations of the radial equation are suitable to JWKB treatment but are inconvenient for integration at small values of r. We combine the methods of Press & Teukolsky and Chandrasekhar to utilize the best features of each. The result is an efficient numerical solution of the radial equation.

7.2.1 The method of Press & Teukolsky and the JWKB approximation

The asymptotic radial equation solutions are of the form

$$_sR \sim K^{\mathscr{H}} \Delta^{-s} e^{-ikr^*} \quad r^* \to -\infty \tag{7.13}$$

(where $K^{\mathscr{H}}$ is a constant) which satisfies the requirement of ingoing flux at the horizon, and

$$_{-2}R \sim K^{\text{in}} \frac{e^{-i\omega r^*}}{r} + K^{\text{out}} r^3 e^{i\omega r^*} \quad r^* \to +\infty \tag{7.14a}$$

(for Ψ_4) and

$$_{+2}R \sim (\tilde{K}^{\text{in}}/r) e^{-i\omega r^*} + (\tilde{K}^{\text{out}}/r^5) e^{i\omega r^*} \tag{7.14b}$$

(for Ψ_0), where K^{in} and $\tilde{K}^{\text{in}}$ are constants, $k = \omega - ma/2Mr_+$, and r^* is the 'tortoise coordinate' defined by (2.64). The practical difficulty in integration of R arises because the computer integrates the sum of the ingoing and outgoing parts of the solution. Due to the r^{2s} peeling factor, for larger r the ingoing piece on Ψ_4 is swamped by the outgoing part, precluding extraction of the separate ingoing and outgoing amplitudes. Press & Teukolsky (1973) devised a technique for transforming such equations which alleviates this problem. However their approach is still susceptible, as is the original equation, to the further difficulty that for larger l-modes integration must proceed to large values of r at such small step size that the computation becomes impractical. We discuss this difficulty below.

In applying the technique of Press & Teukolsky to $_{-2}R$ one makes the substitution $\chi = (R/\tilde{R})'$ where $\tilde{R} = r^3 \exp(i\omega r^* + C_1/r + C_2/r^2 + C_3/r^3)$ is the asymptotic outgoing solution up to and including the C_2/r^2 term, but C_3 is adjustable and does not necessarily correspond to the asymptotic solution (see below) and the prime denotes derivation with respect to r^*. Because of the almost exact cancellation between $\tilde{R}$ and the outgoing solution R, the dominant outgoing piece arises from differentiation of r^{-3}. The dominant ingoing piece arises from differentiation of the phase. Hence the resulting dominant contributions of both ingoing and outgoing parts are of the same order r^{-4} in χ.

If the radial equation (2.27) is written in terms of r^* (cf (2.64)) as

$$R'' = AR' + BR \tag{7.15}$$

then the explicit transformation is to

$$\chi'' = \mathscr{A}\chi' + \mathscr{B}\chi \tag{7.16}$$

with

$$\mathscr{A} = \alpha + \beta'/\beta, \tag{7.17a}$$

$$\mathscr{B} = \alpha' + \beta - \alpha\beta'/\beta, \tag{7.17b}$$

where

$$\alpha = A - 2\tilde{R}'/\tilde{R}, \tag{7.17c}$$

$$\beta = B + A\tilde{R}'/\tilde{R} - \tilde{R}''/\tilde{R}. \tag{7.17d}$$

Asymptotically χ assumes the form (Matzner & Ryan, 1978)

$$\chi_{\text{out}} = \frac{\mathrm{i}F}{2\omega} r^{-4} \left(1 + \frac{G}{Fr} + \mathrm{O}(r^{-2}) \right)$$

$$\chi_{\text{in}} = -2\mathrm{i}\omega r^{-4} \mathrm{e}^{-2\mathrm{i}\omega r^*} \left(1 - \frac{4M + 2C_1}{r} + \mathrm{O}(r^{-2}) \right). \tag{7.18}$$

The constants C_1 and C_2 are given by

$$C_1 = \mathrm{i}(\lambda + 2am\omega)/2\omega$$

$$C_2 = \frac{1}{4\omega^2}(6a^2\omega^2 - 6am\omega - \lambda + 4a\omega^2 Mm\mathrm{i} + 6\mathrm{i}M\omega).$$

C_3 is adjustable to fix the constant F in (7.18). We use

$$C_3 = \frac{4\mathrm{i}}{\omega^3} \frac{{}_2N^m_{l;0}}{{}_2N^m_{l;\pi}} + \frac{1}{6\omega}[-10\mathrm{i}MC_1 - 12Ma^2\omega + 2\mathrm{i}C_2 - \mathrm{i}C_1{}^2 + 8\omega MC_2$$

$$+ 4aMm - 12\mathrm{i}M^2 - 2m\omega a^3\mathrm{i} - \mathrm{i}(\lambda + 6 + m^2)a^2] \tag{7.19}$$

which gives

$$F = \frac{24}{\omega^2} \frac{{}_2N^m_{l;0}}{{}_2N^m_{l;\pi}}$$

(where the ${}_2N^m_{l;0}$, ${}_2N^m_{l;\pi}$ are limiting forms of the angular functions and are defined in chapter 2) and

$$G = 4C_1C_2 + 12MC_2 - 4MC_1{}^2 - 6Ma^2 - 4M\lambda a^2 - 4\mathrm{i}a^2M\omega C_1$$
$$- 12M^2C_1 - 2a^2C_1 - 6\mathrm{i}a^4\omega + 8ma^3\mathrm{i} + 12\mathrm{i}M\omega C_3. \tag{7.20}$$

For convenience $r^4\chi$ is numerically integrated and the ingoing and outgoing amplitudes are extracted at large values of r via the known asymptotic form of the solutions and their Wronskian.

The gravitational scattering problem suffers the same difficulty as the well known problem in partial wave analysis of the quantum mechanical Coulomb problem (cf Gottfried, 1966). The long range potential implies that the contribution to the cross section of each l-mode decreases only slowly with increasing l, i.e. none of the partial waves 'miss' the effective range of the potential. One must therefore integrate the radial equation up to inconveniently large l to insure the remaining contribution is small. If each such calculation were reasonably short this would be manageable but as l increases one must also integrate further in r for convergence to the asymptotic solution $\tilde{R} = r^3 \exp(\sum C_n r^{-n})$ in the construction of χ. One finds that the terms of the series are of order $(E_l/r\omega)$ so that one needs $r \gg l(l+1)/\omega$ for reasonable accuracy. Thus the range of integration must increase roughly as l^2 whereas the step size must remain small to handle the rapidly oscillating phase.

7.2.1.1 JWKB approximation for radial equation

Hence to proceed in this fashion one would seek to apply some form of JWKB approximation to integrate quickly the long, slowly decreasing tail of the solution. To that end one uses (2.63)–(2.65), to cast the radial equation into the form

$$\mathscr{Y}'' + \xi_{\mathscr{Y}} \mathscr{Y} = 0 \tag{7.21}$$

with $\xi_{\mathscr{Y}}$ asymptotically given by

$$\xi_{\mathscr{Y}} = \omega^2 + \frac{2i\omega s}{r} + O\left(\frac{1}{r^2}\right) \tag{7.22}$$

A similar asymptotic form is obtained for the χ equation. The JWKB solution is (cf Mathews & Walker, 1970)

$$W_{\mathrm{JWKB}} = a_+ W_+ + a_- W_- \tag{7.23}$$

with

$$W_\pm = \xi_{\mathscr{Y}}^{-1/4} \exp\left(\pm i \int \xi_{\mathscr{Y}}^{1/2}\, dr^*\right). \tag{7.24}$$

The constants $a_\pm$ may be determined by matching the JWKB form to an exact solution W at any particular point which gives

$$a_\pm = \pm \tfrac{1}{2} i (W W'_\mp - W' W_\mp). \tag{7.25}$$

Where the right-hand side of this equation is evaluated at the match point. $W_\pm$ are not exact solutions, instead satisfying

$$W''_\pm + (\xi_{\mathscr{Y}} + g) W_\pm = 0$$

with

$$g = \frac{1}{4}\frac{\xi''_{\mathscr{Y}}}{\xi_{\mathscr{Y}}} - \frac{5}{16}\left(\frac{\xi'_{\mathscr{Y}}}{\xi_{\mathscr{Y}}}\right)^2.$$

(Mathews & Walker, 1970). We may view the deviation of the JWKB solution from the exact as a functional variation with r of the 'constants' $a_{\pm}$.

In scattering problems one is usually concerned with the normalized scattering amplitude $a \equiv a_-/a_+$ so that the variation of the coefficients directly measures the expected error in the scattering amplitude. The accumulation of error in a may be estimated by writing

$$a' = \frac{a'_-}{a_+} - \frac{a}{a_+}a'_+$$

and using (7.25) and that W satisfies (7.21) for

$$a'_{\pm} = \pm\tfrac{1}{2}\mathrm{i}(WW''_{\mp} - W''W_{\mp}) = \mp\tfrac{1}{2}\mathrm{i}\,gWW_{\mp}.$$

Now

$$\frac{a'_-}{a_+} = -g\left(\frac{W_+}{W_-}\right)\left(-\mathrm{i}\xi_{\mathscr{Y}}^{1/2} + \frac{1}{4}\frac{\xi'_{\mathscr{Y}}}{\xi_{\mathscr{Y}}} + \frac{W'}{W}\right)^{-1},$$

but using (7.23) and (7.24) we have

$$\frac{W'}{W} = -\frac{1}{4}\frac{\xi'_{\mathscr{Y}}}{\xi_{\mathscr{Y}}} + \mathrm{i}\xi_{\mathscr{Y}}^{1/2}\left(\frac{a_+W_+ - a_-W_-}{a_+W_+ + a_-W_-}\right).$$

The term in parentheses is oscillatory with phase $\sim 2\int \xi_{\mathscr{Y}}^{1/2}\,\mathrm{d}r$ and in general either small or oscillates between $\pm\,\mathrm{i}$. In the generic case we neglect both the bracketed term and $\xi'_{\mathscr{Y}}/\xi_{\mathscr{Y}}$ to estimate the error accumulation by

$$\frac{a'_-}{a_+} \cong \mathrm{i}g\frac{W_+}{W_-}\xi_{\mathscr{Y}}^{-1/2}.$$

Similarly,

$$\frac{a}{a_+}a'_+ \cong -\mathrm{i}ag/\xi_{\mathscr{Y}}^{1/2}$$

resulting in the total error estimate

$$a' \cong \frac{\mathrm{i}g}{\xi_{\mathscr{Y}}^{1/2}}\left(a - \frac{W_+}{W_-}\right). \tag{7.26}$$

For the potential form given by (7.22) the JWKB solutions are asymptotically

$$W_{\pm} \cong \xi_{\mathscr{Y}}^{-1/4}\exp(\pm\,\mathrm{i}\omega r \pm 2\mathrm{i}M\omega\ln r \pm s\ln r) \tag{7.27}$$

so that the ratio of solutions in (7.27) varies as r^{2s}. Error therefore accumulates in a as

$$a' \sim \frac{\mathrm{i}g}{\omega}r^{2s}.$$

Since a $1/r$ potential has $g \sim r^{-3}$ it is apparent that use of a straightforward JWKB approximation requires $s < 1$. One concludes such an approach is inapplicable for the gravitational and electromagnetic cases but could be applied to the neutrino scattering problem

7.2.2 *The Chandrasekhar–Detweiler method and the JWKB approximation*

This inapplicability of the JWKB approach is in fact not intrinsic to the problem but is simply an artifact of the NP formalism. Consider a general radial wave equation of the form

$$\frac{d^2\phi}{dr^2} + \left(\omega^2 + \frac{2\gamma}{r} - \frac{E_l}{r^2} + V_h\right)\phi = 0, \tag{7.28}$$

where $V_h \sim O(r^{-2})$ or higher. By a suitable choice of the parameter γ and V_h and if $\phi = r\psi$ where ψ is the probability wave function, (7.28) is the quantum mechanical Coulomb ($\gamma = Ze^2$, $\hbar^2/\mu = 1$) or the Newtonian ($\gamma = 2m\omega^2$) wave scattering problem. By letting $r^* = r + (\gamma/\omega^2)\ln r$ and

$$y = \left(\frac{dr^*}{dr}\right)^{1/2} \phi = r\left(\frac{dr^*}{dr}\right)^{1/2} \psi \tag{7.29}$$

one finds that y satisfies

$$d^2y/dr^{*2} + (\omega^2 - V_*)y = 0 \tag{7.30}$$

with the potential V_* asymptotically of order r^{-2} or smaller, given by

$$V_* = \frac{E_l + 4\mu^2\omega^2 + V_h r^2}{r^2(1 + 2\mu/r)^2} + \left(1 + \frac{2\mu}{r}\right)^{-3}\left[-\frac{\mu}{r^3} + \frac{3\mu^2}{r^4}\left(1 + \frac{2\mu}{r}\right)^{-1}\right] \tag{7.31}$$

with $\mu = \gamma/2\omega^2$. Clearly any scattering with a real potential which admits plane wave solutions at infinity may be put in short range form by a suitable choice of wave function and coordinates. If one chooses to work directly with (real) metric perturbations instead of the complex quantities of the NP formalism one then expects the Schwarzschild case should possess a choice of variable yielding the form (7.30). Indeed this is the case (cf Chandrasekhar (1975a, b) and references therein). Since the Kerr case is asymptotically Schwarzschild, by analogy one concludes that Kerr should also possess such a form. This is not immediate, due to the use of the complex NP formalism with a resultant potential having, in particular, complex γ (see (2.65)), resulting in a potential which is short range but in a complex variable. The solutions then still differ asymptotically by several powers of r.

Since we know a suitable solution with (at least) real $\gamma(r)$ exists we seek a more general transformation of the form $z(r_*) = g(r)R(r) + h(r)R'(r)$ and solve for g and h subject to the form desired of the equation for z. Such a calculation has been performed by Chandrasekhar & Detweiler (1976), following a study of electromagnetic perturbations of the Kerr spacetime by Detweiler (1976), see also Chandrasekhar (1979b); the result being that z satisfies

$$\left(\frac{\mathrm{d}^2}{\mathrm{d}r_*{}^2} + (\omega^2 - v_z)\right)z = 0 \tag{7.32}$$

where z is related to the Teukolsky functions by

$$K_{+2}z = \frac{\rho^8}{\Delta^2}Q - \mathrm{i}\omega(W - 2\mathrm{i}\omega)Y - \frac{\rho^8}{\Delta^2}(W - 2\mathrm{i}\omega)\frac{\mathrm{d}}{\mathrm{d}r_*}Y, \tag{7.33}$$

where $Y = (\Delta^2/\rho^3)_{+2}R$ and

$$\begin{aligned}
\rho &= r^2 + \alpha^2,\\
\alpha^2 &= a^2 - am/\omega,\\
Q &= (\Delta^2/\rho^8)(F + \beta_2),\\
W &= \frac{1}{F - \beta_2}\left(\frac{\mathrm{d}F}{\mathrm{d}r_*} - \kappa_2\right),\\
K &= \nu(\nu + 2) - 4\omega^2\beta_2 - 2\mathrm{i}\omega\kappa_2,\\
\Delta F &= \nu\rho^4 + 3\rho^2(r^2 - a^2) - 3r^2\Delta \equiv q,\\
\beta_2 &= \pm 3\alpha^2,\\
\kappa_2 &= \pm\{36M^2 - 2\nu[\alpha^2(5\nu + 6) - 12a^2] + 2\beta_2\nu(\nu + 2)\}^{1/2},\\
\nu &= \lambda + 2s,
\end{aligned}$$

and r_* is defined up to a constant by

$$\mathrm{d}r/\mathrm{d}r_* = \Delta/\rho^2. \tag{7.34}$$

Then the potential is given by

$$\begin{aligned}
v_z = \frac{\Delta}{\rho^8}\bigg(q - \frac{\rho^2}{(q - \beta_2\Delta)^2}\{&(q - \beta_2\Delta)[\rho^2\Delta q'' - 2\rho^2 q - 2r(q'\Delta - \Delta' q)]\\
&+ \rho^2(\kappa_2\rho^2 - q' + \beta_2\Delta')(q'\Delta - \Delta' q)\}\bigg)
\end{aligned} \tag{7.35}$$

a prime here denoting $\mathrm{d}/\mathrm{d}r$. Due to the four possible choices of signs for the constants β_2 and κ_2 there are four possible potentials. Chandrasekhar (1975b) has shown that each of the potentials yields the same reflection and transmission coefficients so that for the purpose of cross section computation they are interchangeable.

The detailed behavior of the potential v_z has been discussed at length by Chandrasekhar & Detweiler (1976, 1977), and by Chandrasekhar in a series of papers (1975a, b, 1978, 1983). For our present purpose several of the general features are of importance and are distinguished by the presence or absence of superradiance.

In the nonsuperradiant ($\omega > am/2Mr_+ \equiv \omega_s$) case, $r_+{}^2 + \alpha^2 > 0$ and the potential is well behaved everywhere outside the horizon, but is complex. Asymptotically the potential is real, and has the behavior

$$\begin{aligned} v_z \to [v + 2(s-1)]/r^2 & \quad r \to \infty \\ 0 & \quad r \to r_+ \end{aligned} \tag{7.36}$$

In the superradiant case ($\omega < am/2Mr_+$) the asymptotic behavior is unchanged. However, in this case $r_+{}^2 + \alpha^2 = 2Mr_+ - am/\omega < 0$ so that ρ^2 has a zero outside the horizon at some $r = |\alpha| > r_+$ and the $r_*(r)$ relation is double valued with $r_* \to +\infty$ at both the horizon and as $r \to \infty$.

In any case the appropriate solutions for an incoming plane wave are

$$\begin{aligned} z \to \mathrm{e}^{-\mathrm{i}\omega r_*} + \mathscr{R}\mathrm{e}^{\mathrm{i}\omega r_*} & \quad r \to \infty \\ \mathscr{T}\mathrm{e}^{-\mathrm{i}\omega r_*} & \quad r \to r_+. \end{aligned} \tag{7.37}$$

By the usual Wronskian argument one finds (cf Messiah, 1958) that the conservation of flux is expressed by

$$\mathscr{R}^2 + n\mathscr{T}^2 = 1 \tag{7.38}$$

where $n = W(z, \bar{z})_{r>|\alpha|}/W(z, \bar{z})_{r<|\alpha|}$ and $W(z, \bar{z})$ denotes the Wronskian of z and $\bar{z}$. If one examines the behavior of the Wronskian in the vicinity of $r = |\alpha|$ by using its relation to the Wronskian of the R solutions one finds (Chandrasekhar, 1977) $n = (-1)^{2s-1}$. It is then clear that for integral spin and $\omega < \omega_s$ one has $\mathscr{R}^2 - \mathscr{T}^2 = 1$, and, necessarily, super-radiance. For half integral spin the Wronskian does not change sign and superradiance is disallowed. In some cases an additional singularity in v_z occurs due to zeros of $q - \beta_2\Delta$ but this occurrence has no apparent physical significance (Chandrasekhar & Detweiler, 1976; Detweiler, 1977).

In principle the singularities arising in the potential present no insurmountable obstacle to numerical integration but do require special consideration at the singular points. To avoid such consideration we choose to integrate the χ equation, which is well behaved everywhere, from the horizon through the points of singularity of v_z to a value of r at which the error in the JWKB approximation to the function z, as given by (7.26) with

$W_\pm$ appropriate for z, is small. We then match χ and its derivative to z and its derivative, as discussed below, and apply the JWKB approximation to the z equation to continue the integration to large r.

To match the χ solution to z we make use of the explicit relation between ${}_{+2}R$ and z,

$$ {}_{+2}R = \frac{\rho^{-5}}{\Delta}\{\kappa - [\kappa_2 - \beta_2(W + 2i\omega) - i\omega(W - 2i\omega)]\}z + \frac{\rho^3}{\Delta^2}(W - 2i\omega)z' \tag{7.39}$$

where here $(\,)' \equiv d/dr_*$ (see (7.34)). In fact we integrate the $s = -2$ function ${}_{-2}\chi = ({}_{-2}R/\tilde{R})'$ where the prime denotes differentiation with respect to r^* (as opposed to r_*) so in addition use is made of the fact that if ${}_{+2}R$ is a solution of the Teukolsky equation with $s = +2$ then $\Delta^2{}_{+2}R$ is a solution of the Teukolsky equation with $s = -2$. Then using ${}_{-2}\bar{R} = \Delta^2{}_{+2}R$ in (7.33), inserting (7.35) in the definition for χ^* and twice differentiating we obtain a 2×2 matrix T relating $\bar{\chi}$ and $\bar{\chi}'$ to z and z', namely

$$\begin{pmatrix} {}_{-2}\bar{\chi} \\ {}_{-2}\bar{\chi}' \end{pmatrix} = \mathsf{T}\begin{pmatrix} z \\ z' \end{pmatrix}. \tag{7.40}$$

Explicitly if

$$A_1 \equiv \rho^3(fv_z - i\omega A_2)$$

where

$$fv_z = \frac{\Delta^2}{\rho^8 Q}[K - (\kappa - \beta_2 W - 2i\omega\beta_2)(W - 2i\omega)]$$

and

$$A_2 \equiv \rho^3(W - 2i\omega)$$

then

$$T_{11} = \frac{1}{\tilde{\bar{R}}}\left[A_1' - A_2(\omega^2 - v_z)\frac{dr_*}{dr^*} - A_1\left(\frac{\tilde{\bar{R}}'}{\tilde{\bar{R}}}\right)\right]$$

$$T_{12} = \frac{1}{\tilde{\bar{R}}}\left[A_1\frac{dr_*}{dr^*} + A_2' - \left(\frac{\tilde{\bar{R}}'}{\tilde{\bar{R}}}\right)A_2\right]$$

$$T_{21} = T_{11}' - T_{12}\frac{dr_*}{dr^*}(\omega^2 - V_z)$$

and

$$T_{22} = T_{12}' + T_{11}(dr_*/dr^*).$$

Conversely we have

$$\begin{pmatrix} z \\ z' \end{pmatrix} = \mathsf{T}^{-1}\begin{pmatrix} {}_{-2}\bar{\chi} \\ {}_{-2}\bar{\chi}' \end{pmatrix} \tag{7.41}$$

Care must be taken in the transformation from ${}_{-2}\bar{\chi}$ to z and back to preserve the symmetries of $s \to -s$, in particular

$$ {}_{-s}E_l^m(a\omega) = {}_sE_l^m(a\omega). \tag{7.42}$$

Hence $\lambda_{-s} = \lambda_{+s} + 2s$ and we have simply

$$\nu_{+s} = \lambda_{-s}. \tag{7.43}$$

We next apply a JWKB approximation ((7.23)–(7.24)) to z and z' setting

$$z = a_+ z_+ + a_- z_-$$

$$z_\pm = \xi^{-1/4} \exp\left(\pm \mathrm{i} \int_{r_{*(\text{match})}}^{r_*} \xi^{1/2}\, \mathrm{d}r_* \right). \tag{7.44}$$

In fact the integration is performed for convenience in r^* via $\mathrm{d}r_*\, \mathrm{d}r^* = \rho^2/(r^2 + a^2)$ instead of in r_*. The complex function $\xi^{1/2}$ is then integrated to large r at which the asymptotic functions are obtained via (7.44) and using

$$z'(\text{asymptotic}) = a_+ z'_+ + a_- z'_-.$$

Since the ingoing and outgoing parts of both z and $r^4\bar{\chi}$ are of order constant in r there is no numerical difficulty in extracting the mode amplitudes by means of the inverse transformation T^{-1}.

For small values of $r > r_+$ equation (7.16) is integrated directly by a fourth order Runge–Kutta routine. As in (Matzner & Ryan, 1978), integration is initiated at a value of r^*_{initial} sufficiently close to the horizon (large and negative) to reduce the difference between the power down the hole and the difference of incident and excident flux to several per cent. At r^*_{initial} we use the boundary condition

$$R_{\text{in}} = \exp\left[-ikr^* + 4\left(\frac{(M^2 - a^2)^{1/2}}{r_+{}^2 + a^2} \right) r^* \right] \tag{7.45}$$

with the normalization

$$R_{\text{in}}(r^*_{\text{initial}}) = \tilde{R}(r^*_{\text{initial}}). \tag{7.46}$$

Integration proceeds from r^*_{initial} to large positive values of r^* until the estimated error for this case in the JWKB integration, (7.26), is less than one part in 10^3; at that point the Runge–Kutta solution is matched to the JWKB solution just described.

7.3 The Newtonian analogy revisited

Usually one uses a partial wave analysis when the potential in question is of short range so that the phase shifts must only be calculated for a limited range of l. The remaining phase shifts are then an analytically known form of some type. In the standard application of partial wave scattering the short range of the potential implies that the large l-modes are negligibly affected by the potential and the unscattered free wave form of the solution yields the appropriate phase shift for large l. In the present problem however the r^{-1} part of the potential affects even very large-l modes in the scattering so that we cannot use an analytic unscattered form to obtain the large-l phase shifts. Recall the discussion of chapter 6, showing that the large-l Kerr modes approach not an unscattered form but instead a form which possesses the properties of a wave scattered only by a Newtonian potential. Since the Newtonian form is analytically known this result is just as convenient as if the potential were of short range for we do not really care what form the large-l modes assume. All we require is that they assume any analytical form, relieving us of the necessity of numerically computing them. We then use the form they do assume for the modes we do not calculate.

We put the perturbation equation (7.32) for gravitational waves in the Kerr background in Newtonian form by letting $\phi_z = (\mathrm{d}r/\mathrm{d}r_*)^{1/2} z$. Then in the large r limit we have

$$\frac{\mathrm{d}^2\phi_z}{\mathrm{d}r^2} + (\omega^2 - v_z + \mathrm{O}(r^{-4}))\left(\frac{\mathrm{d}r_*}{\mathrm{d}r}\right)^2 \phi_z = 0.$$

Expanding $\mathrm{d}r_*/\mathrm{d}r$ and using the asymptotic form of v_z,

$$v_z \sim \frac{\nu + s(s-1)}{r^2} + \mathrm{O}(r^{-3})$$

$$= \frac{E_l + a^2\omega^2 - 2a\omega m}{r^2} + \mathrm{O}(r^{-3})$$

gives

$$\frac{\mathrm{d}^2\phi_z}{\mathrm{d}r^2} + \left(\omega^2 + \frac{4M\omega^2}{r} + \frac{12M^2\omega^2 - E_l - a^2\omega^2 + 2a\omega m}{r^2}\right)\phi_z \simeq 0.$$

This is essentially an approximation in $l/\omega \gg 1$ since for fixed ω the classical turning point increases approximately as l.

In the Schwarzschild case the angular eigenvalues are $E_l = l(l+1)$. Now if the equation above is compared to (6.5) we see that they are identical except for the appearance of the term $12M^2\omega^2$ above. If we then define

$$\tilde{l} = -\tfrac{1}{2} + \tfrac{1}{2}[1 - 48M^2\omega^2 + 4l(l+1)]^{1/2}$$

the two equations are identical for large l with $l(l+1)$ replaced by $\tilde{l}(\tilde{l}+1)$. For large l then the Schwarzschild phase shifts are given by (6.9) with l replaced by $\tilde{l}$.

In the Schwarzschild case we could then use as a good approximation to the cross section the expression

$$\frac{d\sigma}{d\Omega} \cong \frac{1}{h^2}\left[\left|\sum_P\left(\sum_{l=2}^{L} \tilde{k}_{l2\omega P} + \sum_{l=L+1}^{\infty} \tilde{k}^{\text{Newt}}_{l2\omega P}\right) {}_{-2}S_l{}^2(\theta;0)\right|^2 + \left|\sum_P \sum_{l=2}^{L} (-1)^l P\tilde{k}_{l2\omega P}\, {}_{-2}S_l{}^2(\pi-\theta;0)\right|^2\right] \tag{7.47}$$

where the $\tilde{k}_{l2\omega P}$, defined as in (5.17), are numerically computed using the exact Schwarzschild radial equation and the $\tilde{k}^{\text{Newt}}_{l2\omega P}$ are obtained from (6.9) with the substitution $l \to \tilde{l}$. In fact the summed Newtonian cross section $\sum_{l=2}^{\infty} k^{\text{Newt}}_{l2\omega P}\, {}_{-2}S_l{}^2(\theta;0)$ is known; it is given in (6.44). So we add to (7.47) the summed Newtonian cross section and subtract the Newtonian contributions for $l \leqslant L$. We thus replace the first term in (7.47) by

$$\frac{1}{h^2}\left|\sum_P \sum_{l=2}\left(\tilde{k}_{l2\omega P} - e^{-i\tilde{\alpha}}\frac{h[4\pi(2l+1)]^{1/2}}{2\omega} e^{i\eta_l}\sin\eta_l\right) {}_{-2}S_l{}^2(\theta;0) + hg(\theta)e^{i\tilde{\alpha}}\right|^2 \tag{7.48}$$

where $g(\theta)$ is the summed Newtonian scattering amplitude and $\tilde{\alpha}$ is a constant necessary to reconcile the irrelevant phase difference between r^* and r^*_c.

For $a\omega \neq 0$ one still finds that the phase shifts approach the Newtonian values. If l is large then the perturbing term $\mathscr{L}_1$ in (7.4) is still small relative to $l(l+1)$ and a standard perturbation calculation finds (cf Merzbacher, (1971); Breuer *et al.* (1977))

$$E_l = l(l+1) - \frac{ms^2}{l(l+1)} - \frac{a^2\omega^2}{2} + a\omega \sum_{n=1}^{\infty} P_l(n), \tag{7.49}$$

where $P_l(n)$ are successively higher order corrections to the eigenvalues. Each $P_l(n)$ in the perturbation expansion is a sum of terms sharply peaked around a maximal term

$$P_l(n)_{\max} \sim \left(\frac{\langle Y_{lm}|\mathscr{L}_1|Y_{l+1m}\rangle}{E_l - E_{l+1}}\right)^n.$$

Since $\langle Y_l|\mathscr{L}_1|Y_{l+1}\rangle$ is of order $a^2\omega^2$ times an integral of order unity and

$E_l \sim l(l+1)$ the terms $P_l(n)$ are of order

$$P_l(n) \sim \left(\frac{a^2\omega^2}{l}\right)^n.$$

Hence the condition for rapid convergence of the expansion to the Newtonian values is the requirement familiar from ordinary quantum mechanics that

$$1 \gg \frac{\langle Y_{lm}|\mathscr{L}_1|Y_{km}\rangle}{E_l - E_{k\neq l}} \sim \frac{(a\omega)^2}{l}.$$

A similar condition holds for the ${}_sS_l^m$ themselves, except that the convergence is one order faster in $a^2\omega^2/l$.

The result is that for $a\omega$ fixed the phase shifts still converge to the Newtonian values as l increases. Hence in all cases there exists a value L for which the expression (7.47) provides a good approximation to the cross section. For $a\omega \leqslant 2$, $L \sim 20$ provides accuracy commensurate to the accuracy of the numerical integrations.

8
Absorption, phase shifts and cross sections for gravitational waves

8.1 Introduction

We now examine the details seen in the calculated quantities. Our aim is to supply a physically intuitive context in which to understand the scattering phenomenon as a whole. To that end we first discuss two simplified physically analogous problems, a square barrier and the null torpedo model. Most of the features of the calculated cross sections may be understood in terms of these simplified models.

We discuss in turn the calculated absorption of the incident wave as a function of l, the phase shifts as a function of l and the summed angular cross sections. The absorption as a function of l provides us with a measure of the apparent size of the hole as measured by the marginally trapped null trajectories for each incident mode. Further, by excluding the absorbed modes from the scattering cross section, we find we may anticipate certain features in the cross sections. The l modes which are not absorbed are summed in the angular cross section but contribute with their respective phase shifts. By examining the phase shifts we may understand the features of the angular cross sections in terms of interference of the l modes, governed by their phases. Finally, using physical intuition, we examine the detailed angular cross sections and find several interesting interference phenomena. These phenomena are analogous to similar phenomena seen in numerous classical and quantum scattering processes, in particular the glory phenomena described in section 6. Most of our results are for the scattering of gravitational radiation. However, we also include for comparison some very interesting results on the scattering of scalar waves in the Schwarzschild background, due to Sanchez (1978a, b).

8.2 The square barrier and null torpedoes

We may study the solution of the large ω limit of the radial equation in the form of (7.32) by appealing to an analogous square barrier problem. The potential v_z in the perturbation equation (7.32) approaches zero as $\exp(r_*)$

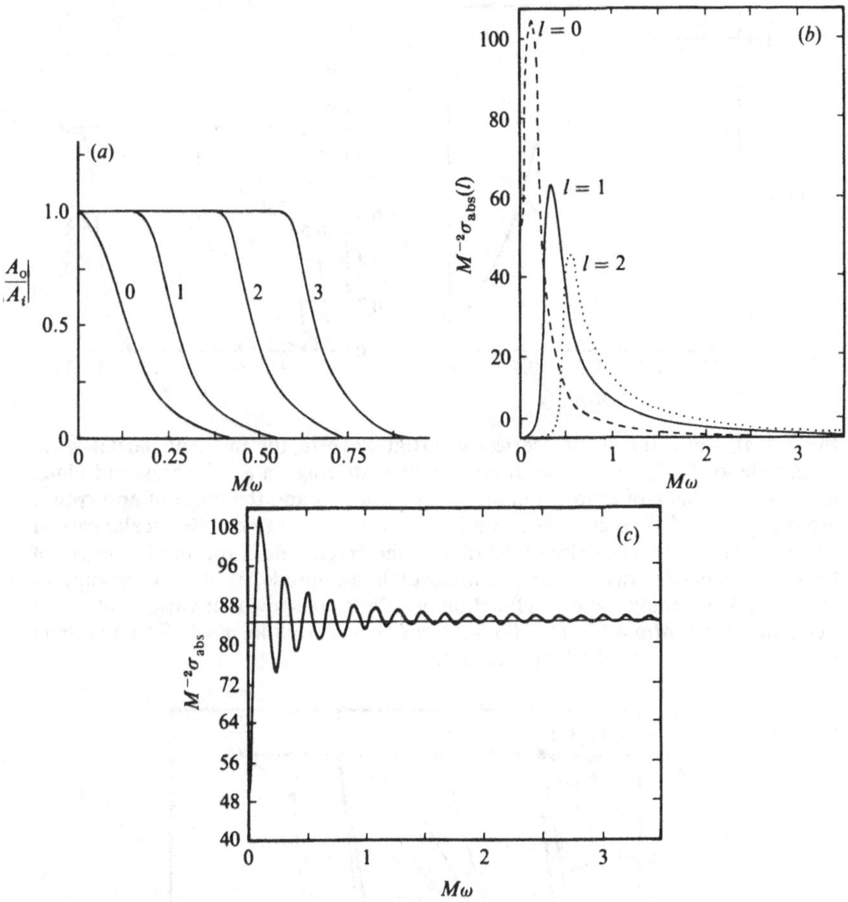

Fig. 8.1. (*a*) The ratio of the outgoing to the incident wave amplitude calculated by Sanchez (1978a) (but the figure was separately prepared by the authors of this work) for scalar waves incident on the Schwarzschild black hole. Curves are labelled by their value of l. Notice that even though $|A_o/A_i| = 1$ at $M\omega = 0$ for $l = 0$, there is in fact a finite partial wave $l = 0$ absorption at zero frequency, and this leads to the finite total absorption cross section, $T_{abs} = 4\pi(2M)^2$ for scalar waves at $\omega = 0$ (see section 6.1.1). (*b*) (After Sanchez 1978a). If any scalar partial wave is partially absorbed, it will contribute a partial cross section equal to

$$4\pi\frac{(2l+1)}{\omega^2}\left(1 - \left|\frac{A_o}{A_i}\right|\right)^2$$

to the total absorption cross section (see section 6.11). The contribution of the first three l modes is plotted. We would expect the total cross section to have a relative maximum near the maximum of each of these partial wave contributions (See fig. 8.1(c).) (*c*) The total absorption cross section for scalar waves as a function of frequency (Sanchez 1978a). At $M\omega = 0$, the absorption cross section is totally given by the $l = 0$ mode and equals $4\pi(2M)^2$. Each partial mode gives a relative maximum in the absorption; the absorption approaches the geometrical optics value $27\pi M^2$ for large $M\omega$. Sanchez (1978a) has shown that the period of the oscillations is $\Delta(M\omega) \cong 0.19$, and that the amplitude of the oscillation decreases as $(\sqrt{2}M\omega)^{-1}$.

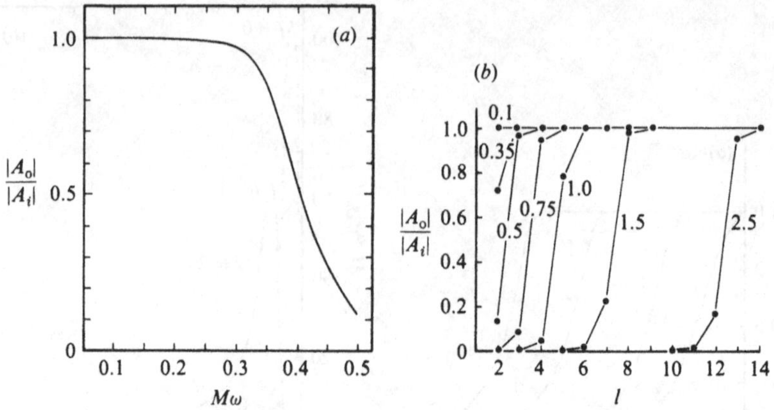

Fig. 8.2.(*a*) The ratio of the scattered partial wave to the incident partial wave amplitude for $l = 2, a = 0$ gravitational wave scattering on a Schwarzschild black hole. Note the lack of absorption for small frequency and the onset of absorption around $M\omega \sim 0.3$. This curve is qualitatively quite similar to the $l = 2$ scalar case of Figure 8.1(a). (*b*) This is a plot of the ratio of the magnitude of the amplitude, A_o, of the scattered partial wave to the magnitude of the amplitude, A_i, of the same angular mode of the incoming wave as a function of l. Plots are shown for various values of frequency from $M\omega = 0.1$ to $M\omega = 2.5$ for Schwarzschild $a = 0$. The numbers labelling each plot are the frequencies $M\omega$.

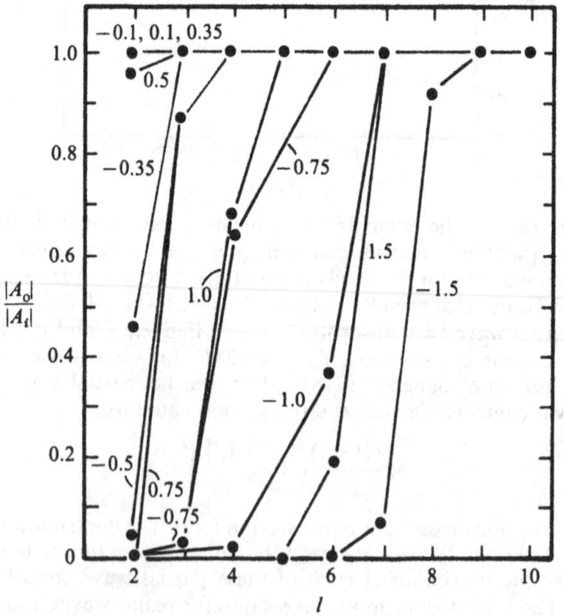

Fig. 8.3. This is a plot similar to fig. 8.2(*b*) of the ratio of the magnitudes of the outgoing scattered to incoming amplitudes, but for gravitational radiation scattering off a Kerr hole with angular momentum given by $a = 0.75M$. The plots are again labelled by frequency $M\omega$. Negative frequency corresponds to counter-rotation, positive to co-rotation. For $a = 0.75M$ superradiance is too small to be visible on this plot.

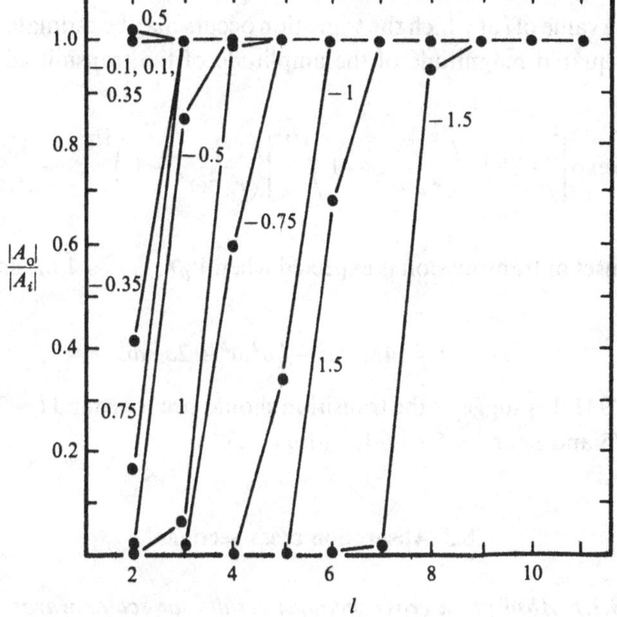

Fig. 8.4. This is a plot similar to figs. 8.2(*b*) and 8.3 of the ratio of the magnitudes of the outgoing scattered to incoming gravitational wave amplitudes but for a Kerr hole with angular momentum given by a $= 0.9M$. Notice the slight superradiance for $M\omega = 0.5$.

as r_* becomes negative and varies as $V_l r_*^{-2} \sim V_l r^{-2}$ for $r_* > 0$, where

$$V_l = E_l - 12M^2\omega^2 + a^2\omega^2 - am\omega. \tag{8.1}$$

This behavior may be approximated by a one-dimensional rectangular barrier in x by a potential of the form

$$\begin{aligned} V(x) &= V_l/r_+{}^2, \quad -b/2 < x < b/2 \\ &= 0, \qquad \text{elsewhere,} \end{aligned} \tag{8.2}$$

for $-\infty < x < \infty$, where b is the classical turning point of the potential of (8.1), $b = V_l^{1/2}/\omega$.

Examining this simplified problem one finds intuitive explanations of many of the features of figs. 8.1–8.4. For small l the energy of the wave $\sim \omega^2$ exceeds the height of the barrier and absorption is essentially complete so the ratio out/in is zero. As l increases the height of the barrier increases until it is comparable to the energy of the incoming wave and the transition is made to complete reflection. Since the simplified problem (8.1) is exactly soluble the value of l at which the transition occurs may be estimated. In this case the squared magnitude of the amplitude of the transmitted wave is given by

$$T^2 \approx 16\exp\left[-2V_l^{1/2}\left(\frac{V_l}{r_+{}^2\omega^2}-1\right)^{1/2}\right]\left(\frac{V_l}{r_+{}^2\omega^2}-1\right)^{1/2}, \quad \frac{V_l}{r_+{}^2\omega^2} \gtrsim 1, \tag{8.3}$$

and the onset of transmission is expected when $V_l/r_+{}^2\omega^2 \sim 1$ or, using (8.1) and (7.49)

$$l(l+1) \sim 14M^2\omega^2 - \tfrac{1}{2}a^2\omega^2 + 2am\omega. \tag{8.4}$$

For $a = 0.9M$ this suggests the transition should occur around $l \sim 2$ or 3 for $M\omega = 0.75$ and near $l \sim 5$ or 6 for $M\omega = 1.5$.

8.3 Absorption cross sections

8.3.1 Absorption cross sections: results for scalar waves

Sanchez (1978a,b) has given a remarkably complete discussion of the scalar case. The radial wave equation was solved by repeated analytic continuation. The partial wave absorption cross section was computed for individual l, analogously to the discussion in section 6.1.1 ((6.11)–(6.15)). Because there is a fairly sharp transition in l between the absorbed and the nonabsorbed modes, each l-mode contributes a 'spike' to the total absorption cross section. Fig. 8.1(a), based on the results from Sanchez (1978a), gives the ratio of the magnitude of the amplitude of the scattered *scalar* partial wave, to the magnitude it has in the plane wave, for the Schwarzschild scalar case.

If there were total absorption, each l-mode would contribute to the absorptive cross section as in the calculation in section 6.1.1, i.e.

$$4\pi[(2l+1)/\omega^2];$$

if the absorption is only partial, this is modified by the factor $(1-|A_o/A_i|^2)$. Fig. 8.1(*b*) shows the contribution to the absorption cross section from the first few l modes (using the absorption ratios shown in fig. 8.1(*a*)), and fig. 8.1(*c*) shows the total absorption cross section as a function of frequency (both figures adapted from Sanchez (1977, 1978a)). For $M\omega = 0$, the absorption is totally that due to the $l = 0$ mode, giving $\Gamma_{\text{abs}} = 4\pi(2M)^2$. As each partial mode begins to contribute it causes a relative maximum in the absorption cross section. Sanchez (1978a) has shown that the asymptotic period of the oscillations is $\Delta(M\omega) \cong 0.19$, and that their amplitude falls as $(\sqrt{2}M\omega)^{-1}$; the limiting value of the absorption cross section ($M\omega \to \infty$) is $27\pi M^2$, the geometrical optics value.

8.3.2 *Absorption cross sections: results for gravity waves*

Comparison with the calculated results displayed in figs 8.1–8.5 indicates reasonable agreement with the simplified problem. Further, the rapidity with which the transition occurs is clear since as l increases the transmission of the rectangular barrier decreases as $\exp(-2V_l/r_+\omega) \sim \exp[-2l(l+1)/r_+\omega]$.

Since r_+ decreases for increasing a, the transition is expected to occur with a faster exponential as a increases. Comparison of the $a = 0.9M$ case fig. 8.4 with the Schwarzschild case in fig. 8.2 suggests this is so. Although the approximation of the simple problem to the exact one is somewhat crude, one still expects a rapid transition from transmission to reflection as l increases and a sensitivity to small changes in the height of the potential barrier. These rough expectations are indeed seen in the calculated values, but with expected modifications in detail owing to the more complicated nature of the exact potential.

The superradiance displayed in fig. 8.4 for the low-l modes of the $M\omega = 0.5$, $a = 0.9M$ and that shown in fig. 8.5 for the $M\omega = 0.75$, $a = 0.99M$ case is unexpected from simple classical models. It is characteristic of black hole physics. The superradiance shown, with the small superradiance seen in the $a = 0.9M$, $M\omega = 0.35$ case (not visible in fig. 8.4) is consistent with the conclusions of Teukolsky & Press (1974) that the amount of superradiance increases as $\omega \to \omega_s = am/2Mr_+$ and cuts off abruptly as $\omega = \omega_s$. As later discussion in this chapter will show, this enhancement of the lower-l modes increases the scattering in the backward direction since the low-l modes (low angular momentum) scatter to larger angles.

There is yet another way of viewing the variation of absorption with l. One may view each l mode as a null trajectory incident along the z-axis with an impact parameter b given by

$$b\omega \approx E_l^{1/2} \sim [l(l+1)]^{1/2}. \tag{8.5}$$

The general features will be the same for either photon torpedoes or

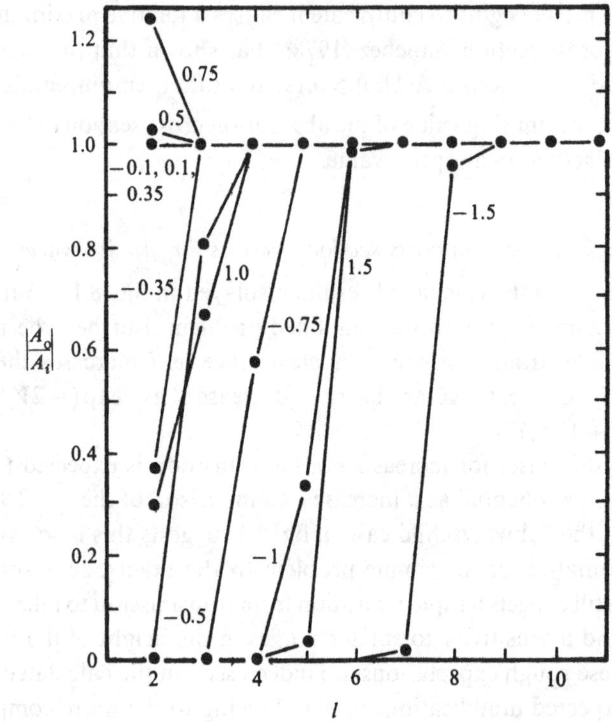

Fig. 8.5. This is a plot of the ratio of scattered to incident gravitational waves like fig. 8.4, but for $a = 0.99M$. Note the enhanced splitting between corotating and counterrotating modes, and the large superradiance for $M\omega = 0.75$.

graviton torpedoes although the details will differ slightly because the potential in the radial equation depends on the spin of the incident flux for finite frequency. The results, except in glory directions, would be identical for all massless fields, in the $\omega \rightarrow \infty$ limit; cf chapter 6. Each torpedo has orbital angular momentum $\sim (E_l)^{1/2} \sim l \sim b\omega$ where b is the asymptotic incident impact parameter $\sim l/\omega$. See fig. 8.6 for an illustration. Notice that the spacing between the incident torpedoes on the b-axis is proportional to $1/\omega$. The energy of each torpedo is $\sim \omega^2$. For b small the torpedo is captured by the hole and we expect complete absorption for such modes. For a mode with just the right value of b, say b_{crit}, the torpedo is marginally captured and orbits the hole indefinitely. For torpedoes with b slightly larger than b_{crit} the mode will spiral the hole and be scattered to large angles. For large b the mode will be scattered in Newtonian fashion and

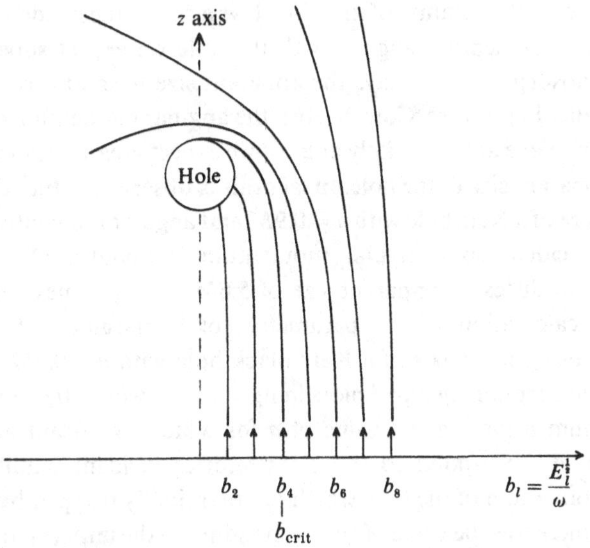

Fig. 8.6. Absorption and scattering analysed in terms of null geodesics: A battery of null torpedoes is shown (arrows at bottom) incident parallel to the z-axis of the hole. Spacing between the torpedoes is $(E_l^{1/2} - E_{l+1}^{1/2})/\omega$. See the text for a discussion.

should contribute to a Newtonian forward divergence.

Consider now the effect of varying the frequency of the incoming wave. At low frequency the energy of the wave is low, the angular momentum barrier in the potential is relatively higher and the spacing in between modes large. At very low frequencies even the first l mode ($l = 2$) has large b and there is no absorption and little scattering to large angles. The cross section should thus have a very Newtonian appearance. For high frequencies the energy is large, the angular momentum barrier is relatively lower and the spacing between l modes is smaller. Many modes have $b \gtrsim b_{\rm crit}$ and we therefore expect much absorption. In addition the close spacing between modes means many modes have b near and slightly larger than $b_{\rm crit}$ and we expect pronounced orbital scattering and scattering to large angles. For very large-l modes we still expect the Newtonian forward divergence.

In this view the l mode at which a transition is made from absorption to reflection has b corresponding to $b_{\rm crit}$. This value of b should serve as a good measure of the apparent size of the hole, corresponding to the impact parameter of a null trajectory which is marginally captured by the hole. In fact the value so obtained in our numerical analysis for the Schwarzschild black hole agrees remarkably well with the analytical value of $b = 3(3)^{1/2}M$.

In the Kerr case the rotation of the hole plays a role; if the incident wave has spin angular momentum aligned with the hole we expect superradiance to inhibit absorption. This says the apparent size from absorption of the hole is smaller in this case. Conversely if the angular momentum of the hole and incident wave are oppositely aligned we expect greater absorption and a larger apparent size of the hole. In fact this is observed in the results. The apparent size of a Kerr hole with $a = 0.9M$ and angular momentum aligned with the incident spin angular momentum is about $3.7M$. Opposite alignment produces an apparent size of $5.6M$. These values constitute a numerical calculation of the parameters of a marginally bound null trajectory along the z-axis of a Kerr black hole with $a = 0.9M$.

For the absorption against l plots in fig. 8.5, one would then expect that the maximum b for a given value of a for which significant absorption occurs would correspond to those trajectories incident parallel to the angular momentum of the hole which are marginally trapped by the hole. Then as ω increases the value of l corresponding to the impact parameter of the marginally bound trajectory would also increase. In fact if the value of l at which absorption begins to cease is taken from fig. 8.2 for Schwarzschild as $l_0 = 3$, 4, and 7 for $M\omega = 0.75$, 0.1 and 1.5 respectively the resultant average apparent size of the Schwarzschild hole is $\sim 5.2M$ which is (probably somewhat coincidentally) the exact result from solution of the geodesic equation $3^{3/2}M = 5.2M$ (Bardeen, 1973).

With this point of view, the split in l values at transition from absorption to reflection in figs. 8.3–8.5 between the corotating and the counterrotating Kerr cases is interesting. If this procedure constitutes a numerical estimation of the apparent size of the Kerr hole as viewed along the rotation axis then using the data in the figures the respective sizes become, for $a = 0.9M$, $b_+ \sim 3.7M$ for the corotating $\omega > 0$ case and $b_- \sim 5.6M$ for the counter-rotating $\omega < 0$ case. This suggests that if one were to solve for the marginally trapped null geodesics with incident direction parallel to the axis of rotation of the hole one should expect a splitting in the apparent size of the hole depending on whether the angular momentum of the hole and spin angular momentum of the incoming wave were aligned or not, with the minimal apparent size corresponding to the case of greatest total angular momentum in the ϕ direction. Indeed, if one estimates the apparent size of the hole from the total absorption cross sections calculated via (5.36) and (5.37) with the mode sums computed here for $a = 0.9M$ ($\sigma = 80.3M^2$ for $M\omega = -1.5$, $\sigma = 62.5M^2$ for $M\omega = +1.5$) one finds $b_+ = 4.46M$ and $b_- = 5.06M$ compared to the numerical Schwarzschild value $5.0M$ for $M\omega = 1.5$. Unpolarized radiation incident axially emerges polarized after interacting with the black hole.

8.4 Phase shifts

Now consider the phases of the scattered partial waves. Figs. 8.7–8.10 display the calculated phase for various ω, and $a = 0$, $a = 0.75M$, $a = 0.9M$ and $a = 0.99M$. For low values of l the phase is uniformly $-\pi/2$. Since the outgoing plane wave amplitude is proportional to i (see (3.78), (3.82)) a phase of $-\pi + \pi/2 = -\pi/2$ indicates total absorption since the corresponding term in the scattering amplitude is then just the negative of the plane wave part. As l increases the phase enters the region of transition between total absorption and reflection and the phases are relatively erratic.

Again this may be understood via the simplified rectangular barrier problem of section 8.2 by solving for the phase in that case in the transition region. For unit incident flux the reflection amplitude for the rectangular barrier is, with $\kappa b = (V_l/r_+{}^2\omega^2 - 1)^{1/2} V_l{}^{1/2}$ small

$$R \approx \exp(-\mathrm{i}V_l^{1/2})\kappa b. \tag{8.6}$$

Hence the phase is $\sim V_l^{1/2} \sim [(l+1)]^{1/2}$. Since the phase is modulo π this leads to the erratic behavior of the phases in the transition region in the figures. It should be remembered that the magnitude of the outgoing wave in these cases is still small.

As l continues to increase into the region of greater reflection the convergence to the Coulomb values becomes evident. The convergence is slower in l the larger the values of $a\omega$ indicating that even for the relatively small value $M\omega = 0.75$ one should compute phase shifts for l past 10 (the value to which Matzner & Ryan (1978) were limited) to obtain the details of the cross sections.

The calculations here are extended only up to $a\omega = 1.35$ ($a = 0.9M$, $M\omega = 1.5$) for which the value $L = 20$ is (cf (7.47)) sufficient. It should be observed that this limitation is not imposed by the radial integration, since the integration of the radial equation via the JWKB approximation becomes more efficient as l increases. Instead the limitation was imposed by the necessary solution of the angular equation for the ${}_sS_l^m$ and ${}_sE_l^m$. To compute cross sections for a range of values of $a\omega$ one must perform the integration of the angular continuation equations which is relatively time consuming. On the other hand if one desires only a specific value of $a\omega$, say for a particular situation of astrophysical interest, other techniques may provide the angular functions and eigenvalues more efficiently. In principle the cross section for any desired choice of parameters may be calculated with relative ease.

The importance of the parity dependence of the phase shifts lies in its implication for significant backward scattering. For any value of l a

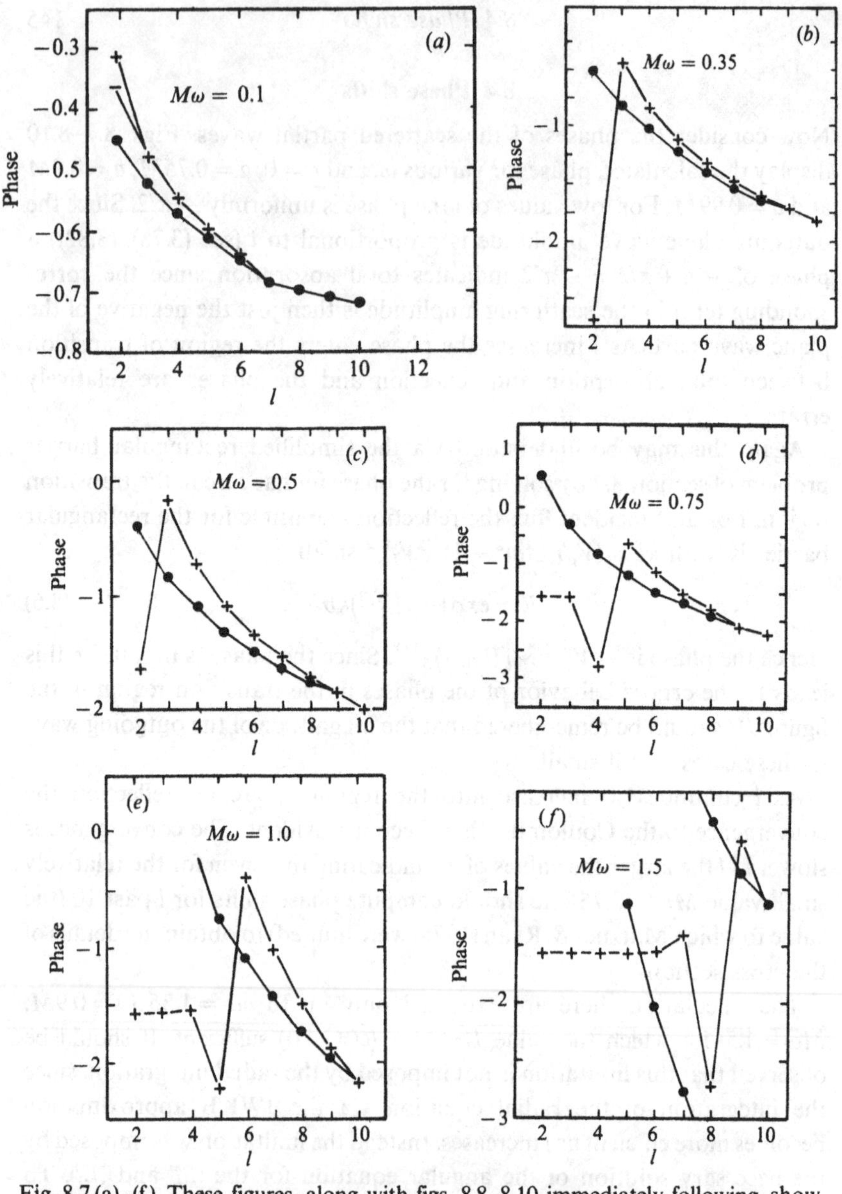

Fig. 8.7.(a)–(f) These figures, along with figs. 8.8–8.10 immediately following, show the phase of the scattered waves for various frequencies for $m = +2$. The $m = -2$ case may be obtained via the reality condition (3.45). The phases of the positive parity $P = +1$ waves are shown with pluses. Minuses indicate the negative parity phases where they are significantly different. The phases from the Newtonian approximation ((6.9) with $l \to \bar{l}$) are shown by dots. As discussed in the text, waves which are completely absorbed possess the phase $-\pi/2$. For the a = 0 cases here, we see absorption for at least some modes when $M\omega = 0.5$ or greater.

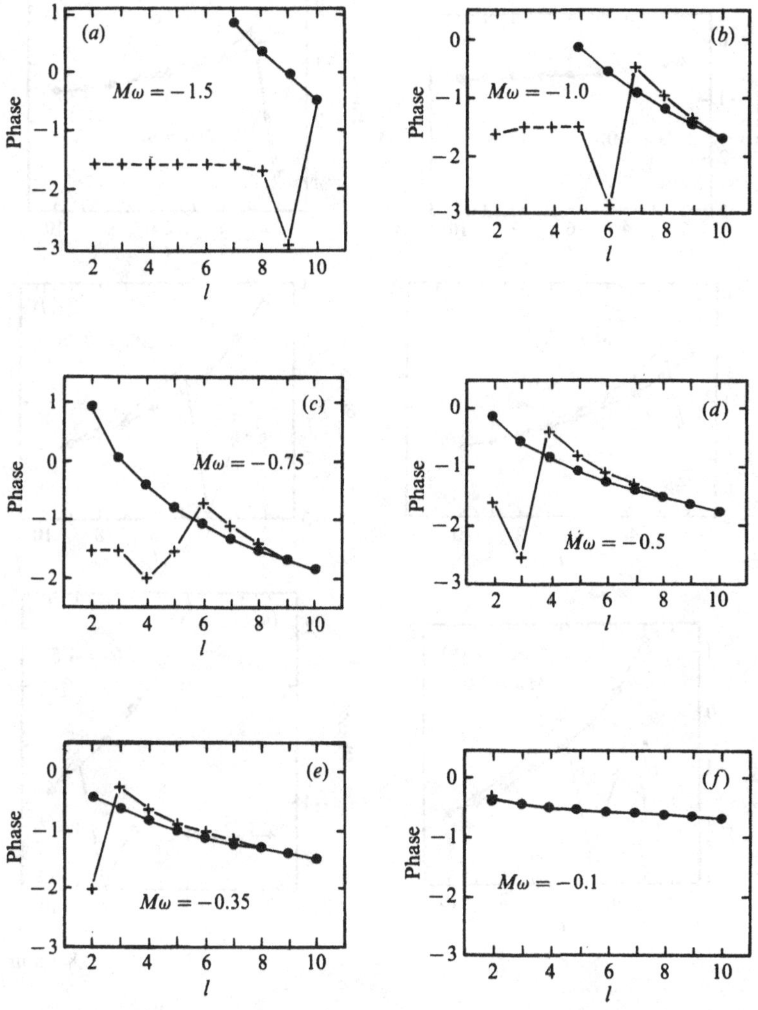

Fig. 8.8. (*a*)–(*l*) Figures similar to 8.7(a)–(f), but for a $= 0.75M$. Because $a \neq 0$, the co-rotating ($M\omega > 0$) and counter-rotating ($M\omega < 0$) cases differ. The counter-rotating frequencies show phases corresponding to total absorption for larger *l*-values than do the co-rotating cases with the same value of $|M\omega|$.

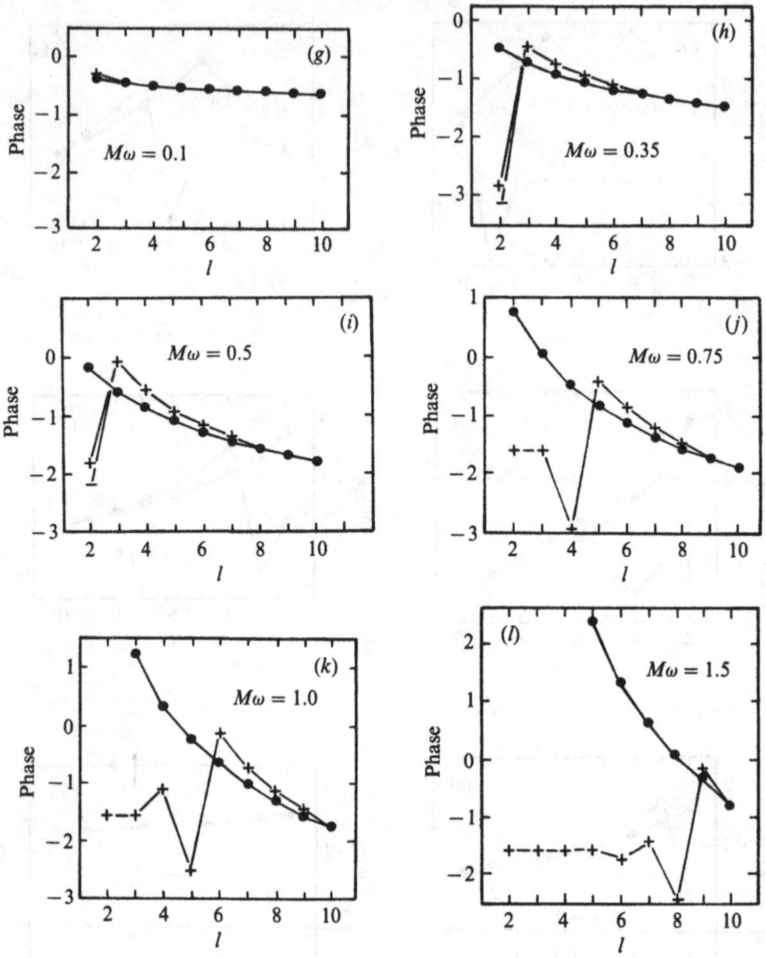

Fig. 8.8 (*cont.*)

backward contribution to the cross section arises only from the second absolute square term in (7.47) since the $_{-2}S_l^2(\theta)$ vanish at $\theta = \pi$ but not at $\theta = 0$ (CMSR). Hence a backward contribution is found only when

$$\tilde{k}_{l2\omega P=1} - \tilde{k}_{l2\omega P=-1} \neq 0, \tag{8.7}$$

in other words, when the difference in phase shifts between the two parities is significant. Now since the radial equation is independent of parity any difference in phase arises from the parity dependence of the asymptotic plane wave $\sim \operatorname{Re} C + 12iM\omega P$. Since $\operatorname{Re} C \sim (E_l)^2$ the phase splitting in parity is maximal for lower values of l but for the effect to contribute to a net phase splitting, the l mode most concerned must not be substantially absorbed (otherwise its phase is pulled by the absorption to $-\pi$). Therefore we expect the maximal phase splitting, hence maximal back scattering, to

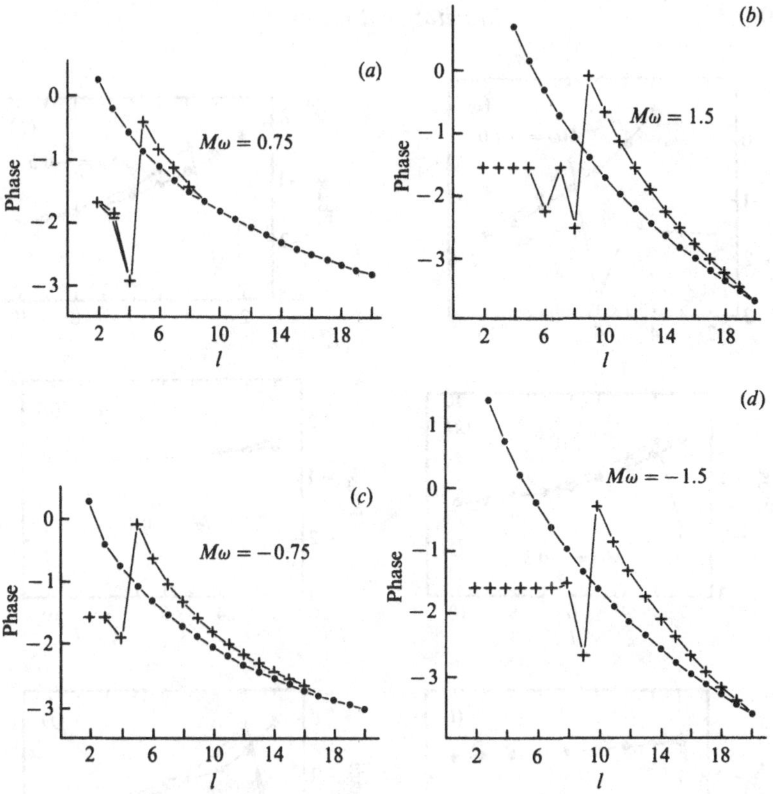

Fig. 8.9.(*a*) For the case shown here, $a = 0.9M$ and corotating frequency $M\omega = 0.75$, the phase for only the lowest $l = 2$ mode shows nearly complete absorption (phase $= -\pi/2$). Between $l = 3$ and $l = 5$ the phase changes rapidly, yielding a large value of the semiclassical deflection function $\Theta(l) = 2\mathrm{d(phase)}/\mathrm{d}l$. (see section 6.2.3). This suggests large angle scattering for these modes. For $l \gtrsim 5$ the Kerr phases approach the Newtonian values rapidly, being virtually identical for $l \gtrsim 8$. (*b*) This plot is similar to fig. 8.9(*a*) but for $M\omega = 1.5$. Notice the larger region for low l (between $l = 2$ and $l = 4$) showing complete absorption. Notice also that the region between complete absorption and approach to the Coulomb values is much larger in l. This, plus the fact that the phases are changing erratically there, leads to a semiclassical expectation of greater interference between modes in the large angle scattering features. The Kerr phases do not become nearly equal to the Coulomb phases until $l \sim 20$; in fact, integration to even higher l-values would be necessary to completely determine the asymptotic l-behavior of the phase. (*c*) This is a plot of the phases similar to fig. 8.9(*a*) but for counter-rotation $M\omega = -0.75$. In comparison to the co-rotating case of fig. 8.9(*a*) notice that this case shows more complete absorption for the low-l modes and only one or two modes which have phases differing greatly from neighboring modes. As discussed in the text this implies less oscillation in the cross sections at large angles since fewer modes participate in large angle scattering, resulting in less interference. Finally notice that the counter-rotating case approaches the Coulomb phases more slowly than the co-rotating case of the same frequency. These features are even more apparent at larger frequency (see fig. 8.7). (*d*) This is the counter-rotating $M\omega = -1.5$ case corresponding to fig. 8.9(*b*). The comments of fig. 8.9(*c*) apply here as well, but this example is even more striking than the lower frequency case. Compare this figure with fig. 8.9(*b*). Fig. 8.9(*e*)–(*l*) $a = 0.9M$ like 8.9(*a*)–8.9(*d*); for various values of $M\omega$.

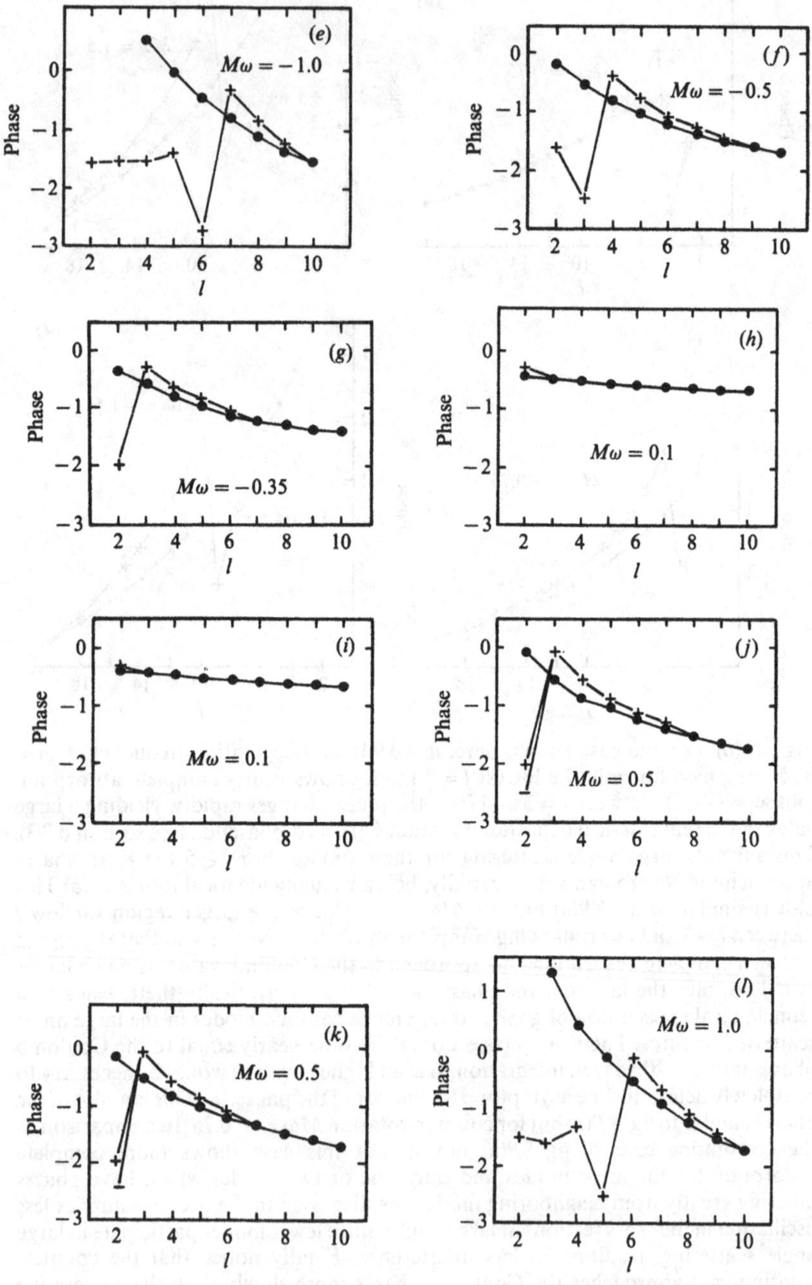

Fig. 8.9 (*cont.*)

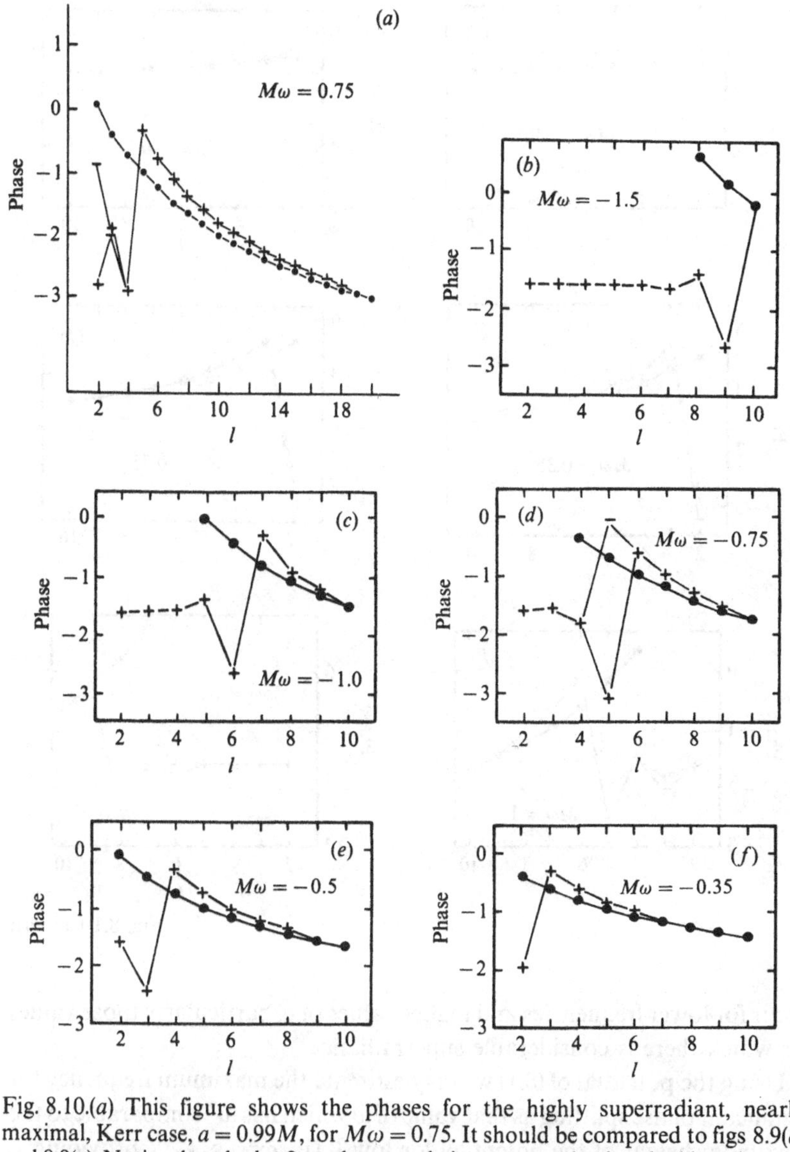

Fig. 8.10.(*a*) This figure shows the phases for the highly superradiant, nearly maximal, Kerr case, $a = 0.99M$, for $M\omega = 0.75$. It should be compared to figs 8.9(*a*) and 8.9(*c*). Notice that the $l = 2$ mode not only is not strongly absorbed but shows a large difference in phase between the positive and negative parity waves. This leads to a large backward contribution as discussed in the text. The intermediate modes, between $l = 3$ and $l = 6$ are very similar to the $a = 0.9M$, $M\omega = 0.75$ case whereas the large-l modes appear similar to the counter-rotating $M\omega = -0.75$ case. Thus we expect a large angle interference pattern similar to that of the counterrotating case, with similar angular behavior of the maxima and minima, but a large background backward contribution due to the superradiance in the lowest two modes. This will be shown in the cross sections in fig. 8.19. Figs. 8.10(*b*)–(*l*) The phases for $a = 0.99M$ and for various frequencies.

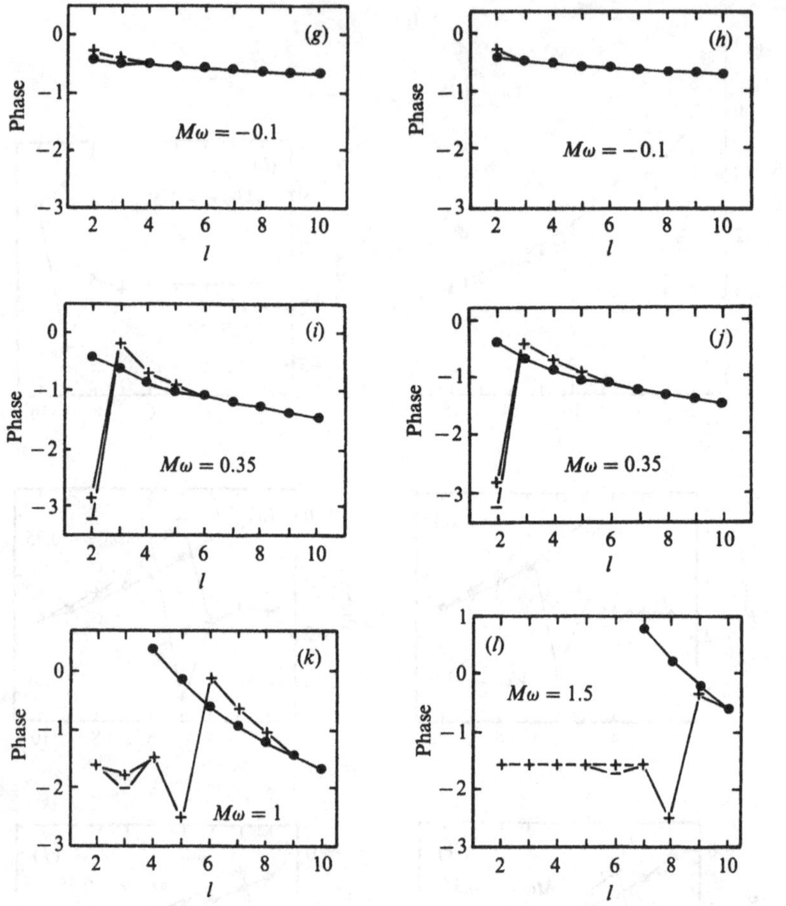

Fig. 8.10 (*cont.*)

occur for lower frequencies and higher values of a, particularly those values for which there is considerable superradiance.

Using the potential of (8.1) we may estimate the maximum frequency for significant phase splitting as that value of ω which has ω^2 comparable to the maximum height of the potential, for low l, i.e., $\omega^2_{\max} \approx V_{l=2}$ providing

$$\omega_{\max} \approx \frac{ma + [m^2a^2 + E_l(12M^2 - a^2 + r_+{}^2)]^{1/2}}{r_+{}^2 + 12M^2 - a^2}. \tag{8.8}$$

For Schwarzschild this gives $M\omega_{\max} \approx 0.6$. For the maximal Kerr case $(a \to M)$, $M\omega_{\max} \approx 0.87$.

Fig. 8.11 illustrates the difference in phases for several cases in which it

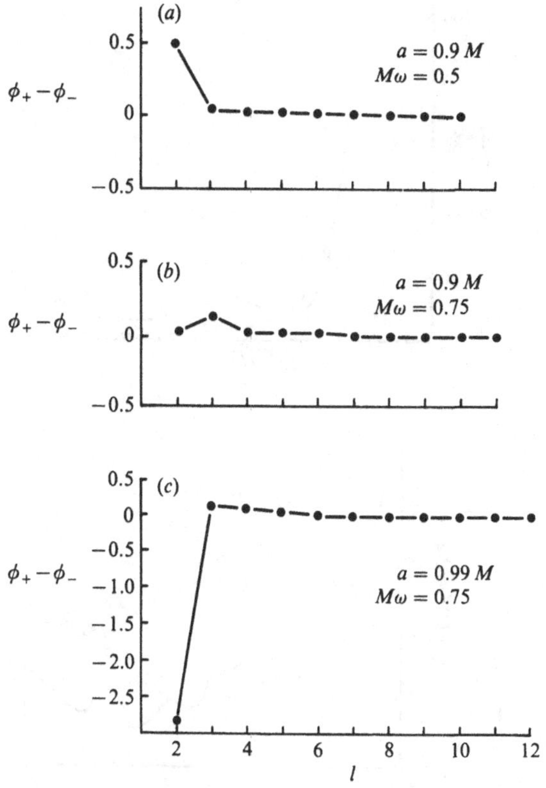

Fig. 8.11. This figure shows the difference $\phi_+ - \phi_-$ between the positive parity $P = +1$ and the negative parity $P = -1$ phases for Kerr holes with several values of frequency $M\omega$ and a. The difference between the phases is important for understanding the backward ($\Theta \sim \pi$) behavior of the cross sections as discussed in the text. These plots are illustrative of the cases in which the difference is significant.

is significant, $a = 0.99M$ and $0.9M$, $M\omega = 0.75$ and 0.5. Frequencies lower than those estimated above also show some phase splitting as pointed out in Matzner & Ryan (1978), but as frequency increases for a given value of a, absorption quickly becomes complete at low l values. The large phase splitting for $a = 0.99$, $M\omega = 0.75$ is particularly enhanced in the resultant cross section since this case shows large (120%) superradiance in the $l = 2$ mode (see fig. 8.5). The resultant significant backward scattering is seen in the angular cross sections plotted below.

Since the parity splitting in phase appears explicitly in the plane wave expression the l dependence of the resultant splitting may be determined. Performing the sum over parity in the second term of (7.47), and using the

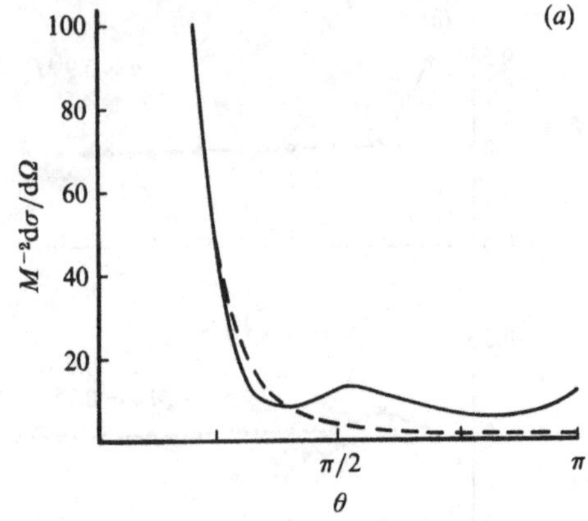

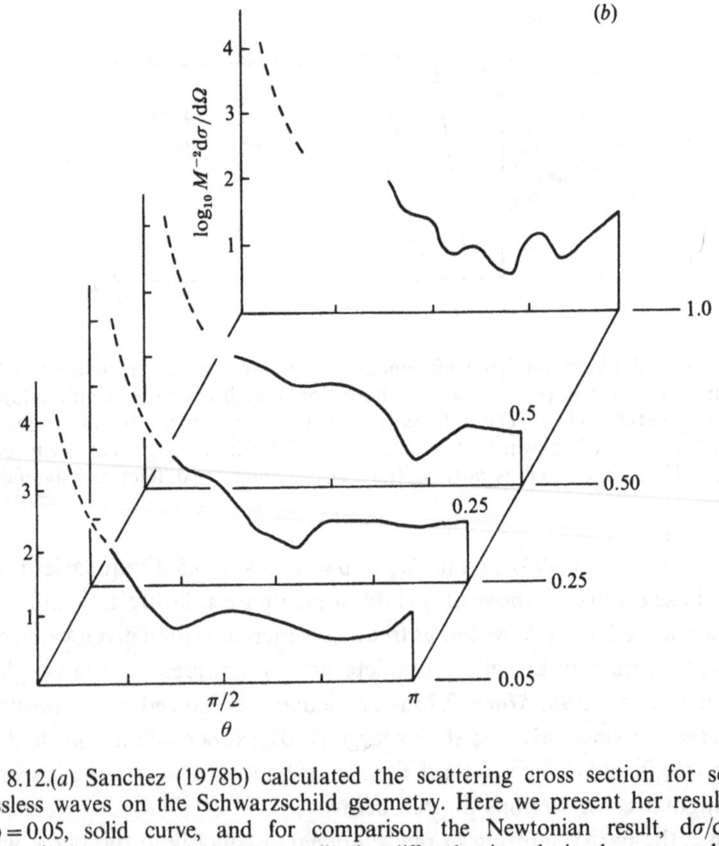

Fig. 8.12.(*a*) Sanchez (1978b) calculated the scattering cross section for scalar massless waves on the Schwarzschild geometry. Here we present her result for $M\omega = 0.05$, solid curve, and for comparison the Newtonian result, $d\sigma/d\Omega = M^2 \sin^{-4}(\theta/2)$. The partial wave amplitudes differ dominantly in the *s*-wave, but it can be seen that there is substantial interference between the enhanced *s*-waves and the rest of the (Newtonian) partial wave amplitudes. See section 6.1.1. (*b*) This

explicit parity dependence one finds the resultant phase term given by

$$\tilde{k}_{l2\omega P=+1} - \tilde{k}_{l2\omega P=-1} = \tilde{k}_{l2\omega P=+1} \frac{24\mathrm{i}M\omega}{[(\mathrm{Re}C)^2 + 144M^2\omega^2]^{1/2}} \times \exp\left[-\mathrm{i}\tan^{-1}\left(\frac{12M\omega}{\mathrm{Re}C}\right)\right]. \tag{8.9}$$

For large l, $\tilde{k} \to \tilde{k}^{\mathrm{Newt}}$ and

$$\mathrm{Re}C \to \mathrm{Re}C(\mathrm{Schwarzschild}) = l(l+1)(l+2)(l-1) \gg 12M\omega; \tag{8.10}$$

hence for large l the phase split term is

$$\sum_P \tilde{k}_{l2\omega P} \xrightarrow[l\ \mathrm{large}]{} \tilde{k}^{\mathrm{Newt}} \frac{24\mathrm{i}M\omega}{l(l+1)(l+2)(l+1)} \propto \tilde{k}^{\mathrm{Newt}} l^{-4}. \tag{8.11}$$

For $l \gtrsim 20$ the relative contribution to the scattering from the large-l modes is less than one part in 10^6 and virtually no contribution to the cross section is lost by truncating the computations at $L = 20$.

8.5 Scattering cross sections

8.5.1 The Schwarzschild scattering of massless scalar waves

Before presenting the results for gravitational wave scattering on black holes, we present here the cross section computed by Sanchez (1978b) for scalar waves scattered by a Schwarzschild black hole. For low frequency the scattering amplitude is, as noted in section 6, essentially that for Newtonian (Coulomb) scattering, except for the large s-wave scattering. Interference between the s-wave and other waves can be expected to lead to significant oscillation in the scalar cross section, even for quite low frequency. In fig. 8.12(*a*) we plot the $M\omega = 0.05$ scalar cross section computed by Sanchez (1978b), and for comparison, the Newtonian cross section $M^2/\sin^4(\theta/2)$. The oscillations are obvious. Fig. 8.12(*b*) plots (from Sanchez 1978b) the scattering cross section for a number of different values of $M\omega$.

Caption for fig. 8.12 (*Cont.*)
graph is drawn from plots given by Sanchez (1978b). Here we plot the logarithm (base 10) of the cross section for various values of $M\omega$, computed for scalar massless waves scattering on the Schwarzschild structure. In each case the dotted extension near $\theta = 0$ is the (logarithm of the) Newtonian cross section (the dotted curve in fig. 8.12(*a*)). For low frequencies this matches smoothly, but especially for $M\omega = 1.0$ there is some structure near $\theta = \pi/4$ not given in Sanchez's results, which would have to be known to fit the cross section onto the Newtonian forward peak which we know it to follow near the forward direction.

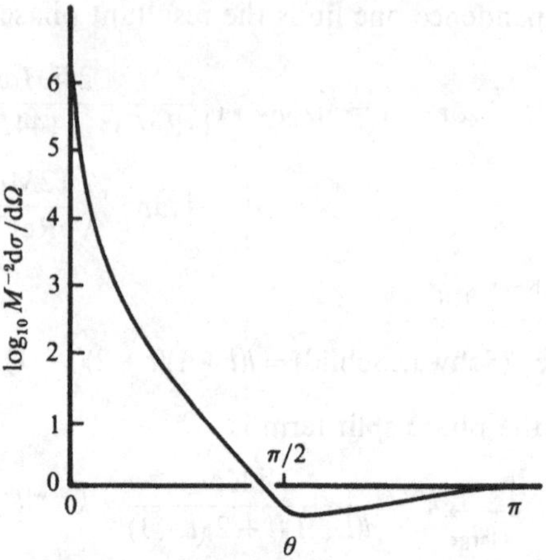

Fig. 8.13. The differential cross section as a function of angle for Schwarzschild with $M\omega = 0.1$, $a = 0$ (Matzner & Ryan, 1978). This plot presents the low frequency scattering cross section. Notice the Coulomb like behavior at all angles, especially the lack of significant backward scattering. This is due to nearly complete absorption of all low-l modes. Compare this plot with the higher frequency cross sections in fig. 8.14.

8.5.2 The Schwarzschild scattering of gravitational waves

We now turn our attention to the calculated angular cross sections themselves and find a wealth of interesting structure. The cross sections chosen for illustration are presented in figs. 8.13–8.17 and 8.19–8.20. Figs. 8.13 and 8.14 display the Schwarzschild cross sections for various values of $M\omega$. These serve as signposts to understanding the considerably more intricate structure in the Kerr cross sections in figs. 8.16, 8.17, 8.19 and 8.20.

In both the Schwarzschild and Kerr cross sections at least three major features are expected: a Newtonian background, a glory, and evidence of orbiting scattering. We now discuss these features in turn.

All the cross sections display the characteristic Newtonian forward divergence and $[\sin(\theta/2)]^{-2}$ fall off near the forward direction and a tendency for small backward scattering. The overall Newtonian background is, however, drastically modified by the other two features (absent from the pure Newtonian case) which evidence behavior modifications peculiar to the relativistic 'pit in the potential' problem (Misner *et al.* 1973).

The detailed angular structure observed in the cross sections has analogs in our simple classical null torpedo model. For large frequencies we should

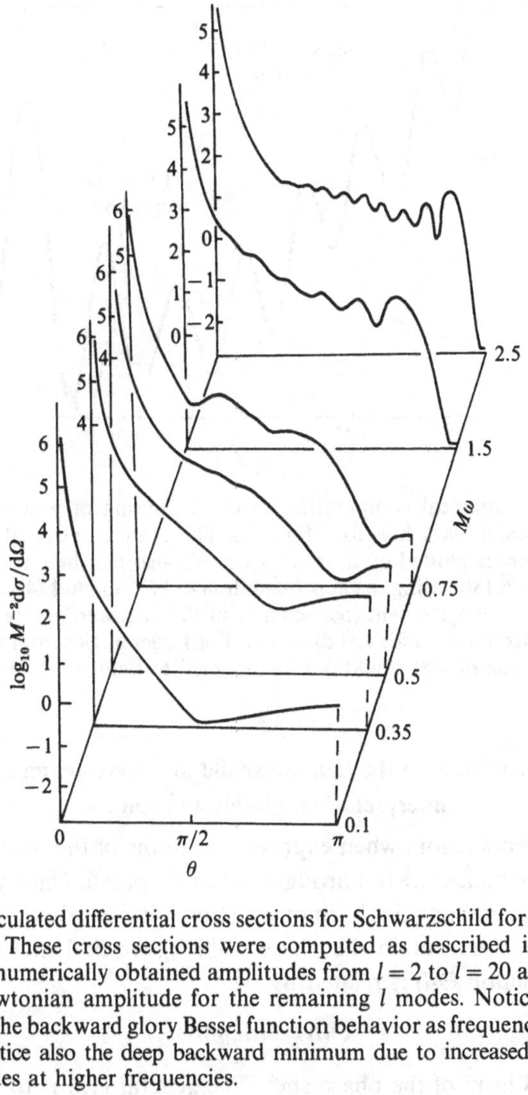

Fig. 8.14. Calculated differential cross sections for Schwarzschild for several values of $M\omega, a = 0$. These cross sections were computed as described in the text by summing the numerically obtained amplitudes from $l = 2$ to $l = 20$ and adding the truncated Newtonian amplitude for the remaining l modes. Notice the striking emergence of the backward glory Bessel function behavior as frequency increases (cf chapter 6). Notice also the deep backward minimum due to increased absorption of the low l-modes at higher frequencies.

certainly have spiral scattering. Such spiral scattering occurs when the photon orbit spirals repeatedly near $r = 3M$, and appears as a characteristic feature in the angular cross sections near the value $\theta \sim \pi/2$. In addition whenever spiral scattering occurs even classically we necessarily have a glory in the backward direction, evidenced by the characteristic Bessel function oscillatory behavior near $\theta = \pi$.

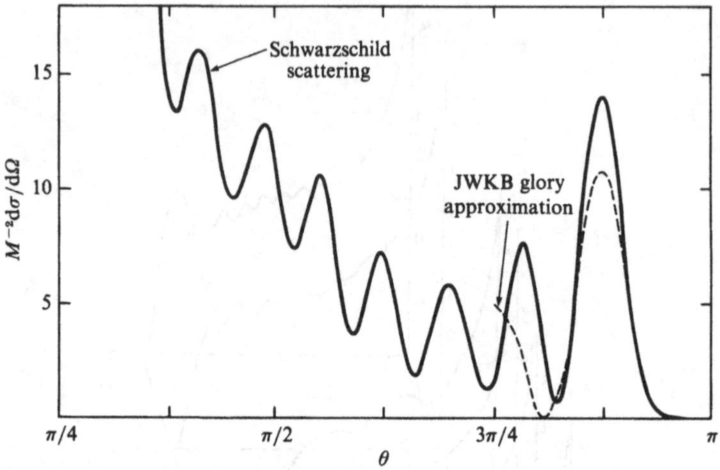

Fig. 8.15. The numerical computation of the scattering of gravitational waves by a Schwarzschild black hole for $M\omega = 2.5$. This is the same result as plotted in fig. 8.14, but here is plotted on a linear scale. We also plot here the JWKB glory approximation (6.188), using the approximation (6.172) and (6.174) (Matzner *et al.*, 1985). We see close agreement (particularly in the backward suppression) for all angles $\gtrsim \pi/10$ from the backward direction. For larger angles from the backward, the approximation (6.188) would not be expected to hold.

8.6 Scattering in the Schwarzschild and Kerr geometries – interpretation: glories and spirals

The glory phenomenon, when expressed in terms of the incident angular momentum, has a feature not brought out in chapter 6. There we avoided a decomposition into angular harmonics. However, if we do carry out such a decomposition and obtain the phase shifts η_l, we find that the classical deflection function $\Theta(l)$ is related by

$$\Theta(l) = 2(\mathrm{d}\eta_l/\mathrm{d}l) \tag{8.12}$$

to the JWKB form of the phase shift. The general idea is to examine the behavior of the classical deflection function $\Theta(l)$ at characteristic values and consider ensuing interference between l modes of nearly the same value of Θ. Since the classical cross section is the product $l(\mathrm{d}l/\mathrm{d}\theta(l))(\sin\theta)^{-1}$, singularities can arise in the classical cross section from the inverse of $\sin\Theta$. If $\Theta(l) = \pi$ and $\mathrm{d}l/\mathrm{d}\theta \neq 0$ for some value of l there is a classical backward divergence. The semiclassical analysis of chapter 6 leads instead to a nondivergent cross section with a Bessel function oscillatory behavior, cf (6.188); note that $l = B\omega$ for these massless fields. In the black

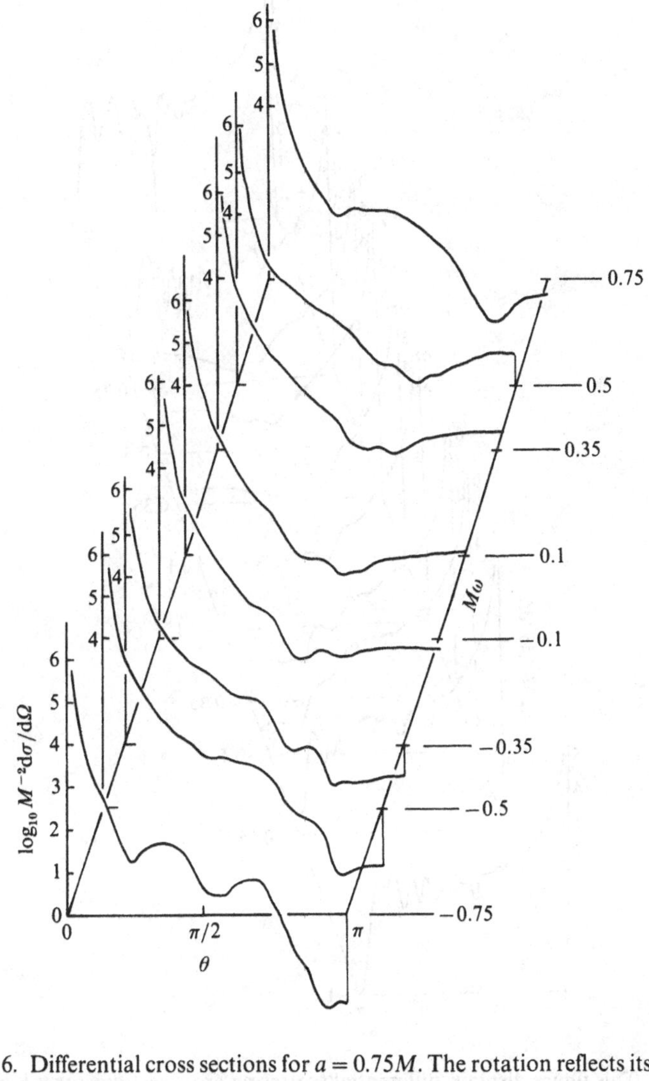

Fig. 8.16. Differential cross sections for $a = 0.75M$. The rotation reflects itself in the different behavior for the $M\omega > 0$ (corotating) and $M\omega < 0$ (counterrotating) cases. It will be noticed that the counter-rotating cross sections more closely resemble the non-rotating (Schwarzschild) cases reported fig. 8.14. The broad backward enhancement here for $M\omega = 0.35$, $M\omega = 0.5$ is due to a small split between the two phases for these frequencies.

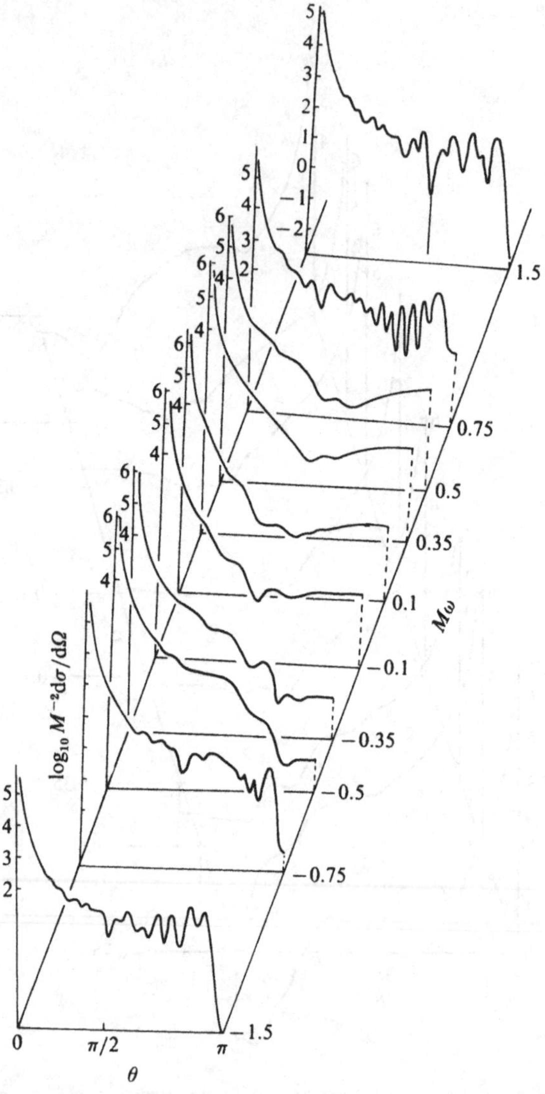

Fig. 8.17. This figure displays differential scattering cross sections for a Kerr hole with $a = 0.9M$ for various values of incident frequency $M\omega$. For $|M\omega| \geqslant 0.75$, each shows the characteristic backward glory near $\theta = \pi$, but notice the enhanced oscillation in the corotating positive $M\omega$ cases due to reduced absorption of the low l-modes. The orbital features are distinct in all cases at angles slightly less than $\theta = \pi/2$ although the smaller dip occurs at slightly lower angle in the corotating case. The source of the peculiar deep narrow dip at $\theta \sim 2\pi/3$ in the $M\omega = 1.5$ case is obscure. The orbital dips seen here should be compared to the generic case shown in fig. 8.18.

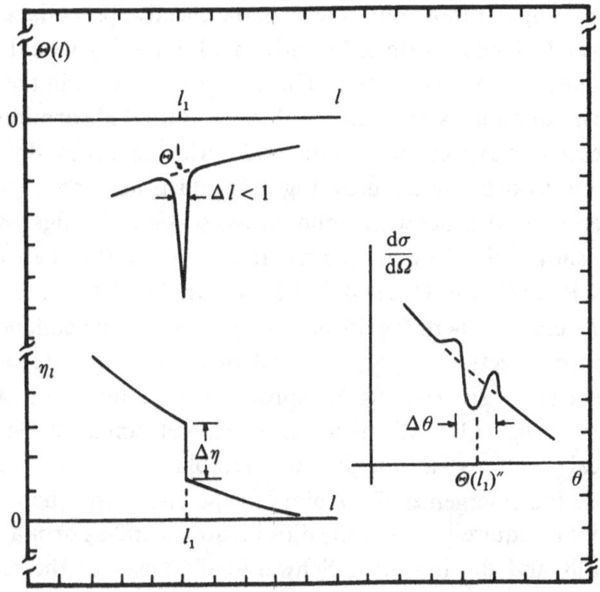

Fig. 8.18. An illustration of a sharp orbital dip from Ford & Wheeler (1959a). The sharp discontinuity $\Delta\eta_l$ in the phase shift η_l at l_1 leads to a deep minimum in the deflection function $\Theta(l)$ at l_1. The characteristic orbital dip then appears at $\theta = \Theta'(l_1)$ where $\Theta'(l)$ is the smoothed deflection function ignoring the discontinuity in η_l. The angular width of the orbital dip is $\delta\theta \sim (\mathrm{d}\Theta'/\mathrm{d}l)_{l_1}^{1/2}$. Compare the characteristic shape of the orbital dip with the features seen in the Kerr cross sections of figs. 8.16, 8.17 and 8.19.

hole cross sections the values of B leading to glories (of which there may be several) are those values immediately following the values which are totally absorbed (see figs. 8.7–8.10). Expectation of these general features is a significant aid in interpreting the calculated cross sections.

Orbiting or spiral scattering occurs when the effective potential possesses a relative maximum equal to the available energy ($\sim \omega^2$) for some value of $l(\sim$ angular momentum). (Again see Ford & Wheeler (1959a) for a definitive analysis.) Intuitively the idea is as follows. We may consider each l mode for a given ω as a projectile impinging on the hole with impact parameter at infinity $B \sim l/\omega$. As l is decreased B will eventually reach the value at which the incoming mode is marginally trapped and infinitely spirals the hole as discussed previously. These modes which spiral the hole will contribute to the cross section at the full range of angles and will appear as a deficit at the angles to which they would otherwise have contributed in the absence of spiralling (e.g, the Newtonian case). This leads to a characteristic orbiting dip (see fig. 8.17 at $\theta = \pi/3$ for an example). In classical potentials the repulsive r^{-2} angular momentum term in the effective potential eventually dominates and all modes with sufficiently small B to spiral would spiral out and contribute to the cross section. The presence of the pit in the potential here however dominates at small r and leads to total absorption for such modes. Hence one expects only those modes which are partially absorbed to contribute to orbiting features. Fig. 8.18 illustrates orbital scattering.

Consider now the Schwarzschild cross sections in figs. 8.13–8.15. Fig. 8.13 shows the low frequency result. $M\omega = 0.1$, calculated by Matzner & Ryan (1978). This should be compared with fig. 6.2, the strictly Newtonian version. There is evidence in fig. 8.13 of some enhancement in the backward direction; absorption and phase splitting are both quite small. The behavior shown in fig. 8.13 proceeds smoothly to the $M\omega = 0.75$ case shown in fig. 8.14, which begins to exhibit some of the expected semiclassical features although considerable intuition is required to recognize the emergence of the glory in the backward quadrant. Near clairvoyance is required to view the dips below $\theta = \pi/2$ as orbital dips (but compare this and the following Schwarzschild cases to the Kerr cross sections which have enhanced orbital dips owing to the increased tendency to orbit the rotating hole). The significantly reduced scattering in the backward direction gives evidence of the increasing absorption of low l-modes that we already saw in fig. 8.2. The total absorption cross section of the hole at this frequency is $72.4M^2$ corresponding to a capture radius of $b = 4.8M$.

The cross sections in fig. 8.14 for $M\omega = 1.5$ and 2.5 give clear evidence of the Bessel function oscillations characteristic of glory scattering. Progressing toward the forward direction from near $\theta = \pi$ the regular glory oscillations give way to the somewhat washed out orbital dip near $\frac{1}{2}\pi$. Notice that as frequency (and hence absorption) increases the depth of the backward cross section minimum increases and the glory crowds closer to π. From the phase shift variation around those values of l which are partially absorbed we expect these cases to have $l_g \sim 8$ and 14 respectively. Indeed the angular separations of the glory minima correspond closely to these values. The calculated total absorption cross sections for the two cases are $83.36M^2$ and $83.61M^2$ respectively corresponding to minimal impact parameters of $b = 5.15M$ and $5.20M$.

By plotting in fig. 8.14 the *logarithm* of the cross section we distort the behavior from a simple-appearing Bessel-like one. In fig. 8.15, we present the backward behavior of the Schwarzschild cross section for $M\omega = 2.5$, plotted on a linear scale. The oscillatory behavior is much more striking. We also, on the same scale present the JWKB approximation of

(6.188), using the approximations of (6.172) and (6.174) (Matzner *et al.*, 1985). We see that there is close agreement in the backward direction, and deviation between the two curves only for angles more than $\sim \pi/10$ away from the backward direction, where the approximations of chapter 6 would surely fail.

The general features seen in the Schwarzschild cross sections are then clear: In the forward direction one finds a Coulomb divergence falling off as $[\sin \frac{1}{2}\theta]^{-4}$ giving way to at least one somewhat subdued orbital dip in the vicinity of $\theta = \pi/2$. If the cross sections for $M\omega = 0.75$, 1.5 and 2.5 are overlaid the orbital dip $\theta \sim \pi/3$ appears to occur at very nearly the same angle. A pronounced glory is seen; its characteristic Bessel function behavior occurs closer to the backward direction as frequency increases. A deep minimum in the backward direction becomes sharper and deeper as frequency increases. Finally, as the frequency increases the total absorption cross section approaches the analytical value $27\pi M^2$.

Fig. 8.16 (from Matzner & Ryan, 1978) gives the Kerr cross sections for $a = 0.75M$, and a range of $M\omega$. However the restriction in that paper to mode sums only up to $L = 10$ prevented extension to higher frequency cases where we would expect to see the semiclassical effect.

Examination of the Kerr cross sections reveals the same general structure with interesting peculiar complications. Consider now the corotating cases in fig. 8.17 for $a = 0.9M$, especially the $M\omega = 0.75$ case. Apart from the Coulomb forward divergence a pronounced orbital dip appears again around $\theta = \frac{1}{3}\pi$ followed by either secondary dips or interference due to the same disturbance.

The angular width of the pronounced dip correlates well with the value of $\delta\theta \sim (\mathrm{d}\Theta_{\mathrm{Coul}}/\mathrm{d}l)^{1/2}$ expected from the analysis of Ford & Wheeler. The glory near $\theta = \pi$ displays a beautiful oscillatory behavior. In the backward direction the minimum is evident but significantly more subdued than the corresponding Schwarzschild case owing to the relatively slight absorption in Kerr at this corotating frequency. The total absorption cross section in this $a = 0.9M$, $M\omega = 0.75$ case is only $36.5M^2$ corresponding to $b = 3.41M$.

Fig. 8.17 also displays the $M\omega = 1.5$ case. The orbiting dip at $\theta \sim \frac{1}{3}\pi$ is significantly modified probably due to an increased tendency for absorption while the dip closer to $\theta = \frac{1}{2}\pi$ appears more prominent in this case. The glory occurs at nearly the same angle as the $M\omega = 0.75$ case but the oscillatory behavior is considerably more complicated due possibly to interference with orbital features. A feature unique to this case is the peculiar extremely deep and narrow dip at $\theta = 0.735$ of width $\delta\theta \sim 0.015$. Its origin is uncertain but it may be an orbital feature corresponding to the

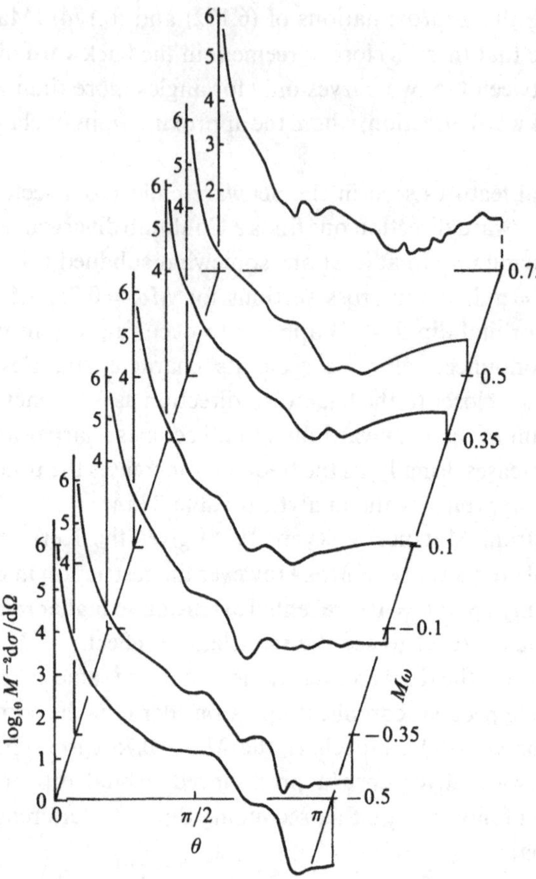

Fig. 8.19. The scattering cross sections for various frequencies for the rapidly rotating black hole with $a = 0.99M$. Notice that there is substantial backward scattering, even for relatively large values of $M\omega$ (due to the superradiance) in the low-l modes. See fig. 8.20.

larger values of l where $d\Theta/dl$ is smaller. The total absorption cross section in this case is $62.5M^2$ corresponding to $b = 4.46M$. This produces the deep minimum in the direct backward direction since the low-l modes which otherwise produce backward scattering are absorbed.

The cross sections for the counter-rotational cases are considerably less oscillatory. This is to be expected from examination of the phases in figs. 8.9 in light of the semiclassical analysis. The counterrotating phases show generally a narrower region in l of greatly varying phase shift. This tends to suppress the 'semi-classical features' apparent in the other cross sections. Fig. 8.17 displays the $a = 0.9M$, $M\omega = -0.75$ case. A pronounced orbiting

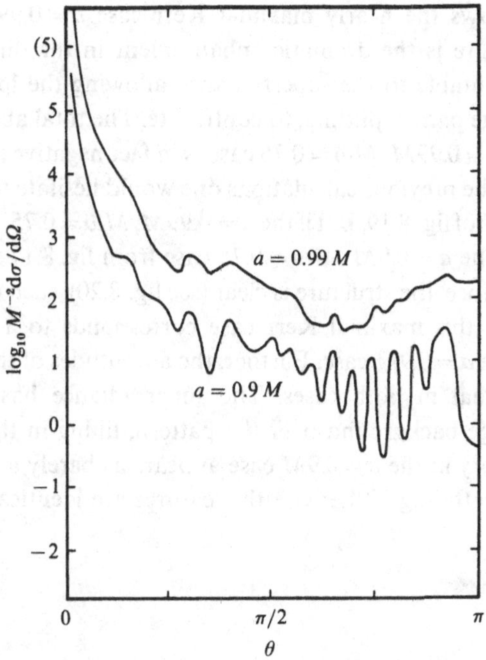

Fig. 8.20. This plot shows two differential cross sections as a function of θ, displaying the effect of superradiance. Both curves are for an incident frequency of $M\omega = 0.75$. The lower curve is for a hole with $a = 0.9M$ for which superradiance is relatively small. The upper curve is for the nearly maximally rotating case $a = 0.99M$ with large superradiance. The latter curve has been displaced upward by a factor of 10(+ 1 on the logarithmic scale) for clarity. The value $\log_{10}(M^{-2}d\sigma/d\Omega) = 5$ in parentheses corresponds to the upper curve only and sets the scale for the plot. Notice the coincidence of the angles at which the maxima and minima occur in the two cases. Notice also that the oscillatory behavior in the backward direction is more pronounced in the non-superradiant case, $a = 0.9M$.

dip is evidenced just below $\frac{1}{2}\pi$ and a secondary spike occurs at $\frac{1}{3}\pi$ where the corotating case possessed a dip. The glory is considerably less glorious in the counterrotating case but comparison to the corotating glory shows the positions of the maxima and minima of the two to be the same but with reduced amplitude. The backward minimum is considerably deeper than the corotating minimum since the total absorption cross section here is $88.7M^2$ corresponding to $b = 5.31$, very nearly the Schwarzschild value.

The forward features of the counterrotating $a = 0.9M$, $M\omega = -1.5$ case in fig. 8.17 are nearly a carbon copy of the corotating case. The glory is considerably less oscillatory and there is no evidence of the sharp spike seen for the corotating case. The backward minimum is virtually zero. In this case the absorption cross section is $80.3M^2$ corresponding to $b = 5.06M$.

Fig. 8.19 shows the nearly maximal Kerr case $a = 0.99M$. The most prominent feature is the dramatic enhancement in the direct backward direction attributable to the superradiance allowing the lowest-l modes, which have large parity splitting, to contribute. The total absorption cross section in the $a = 0.99M$, $M\omega = 0.75$ case is in fact negative ($-15.8M^2$). In the absence of the previous calculations one would hesitate to interpret the angular features of fig. 8.19. But if the $a = 0.99M$, $M\omega = 0.75$ case in fig. 8.19 is overlaid on the $a = 0.9M$, $M\omega = 0.75$ case from fig. 8.17 where there is little superradiance, the structure is clear (see fig. 8.20): each of the maxima and minima in this maximal Kerr case corresponds to a maximum or minimum in the $a = 0.9M$ case. Further, the amplitudes of the maxima are very nearly equal in both cases. The superradiance has the effect of imposing a large background over the pattern, filling in the interference minima; the glory in the $a = 0.9M$ case appears as barely a dimple on the superradiant scattering. Otherwise the features are identical.

9
Conclusion

The analysis of scattering by black holes first led us to consider many different formalisms for the wave perturbations in black hole spacetimes. When we had the perturbations in hand, we came against the problem of defining 'plane waves' in the long range Newtonian tail of the black-hole gravitational field. The analogy with Newtonian gravity gives the solution to this problem, and the solution to the integer-spin case for low frequencies; both embodied in the natural (Regge & Wheeler, 1957) radial variable r^*.

The temptation exists to simply write down a partial sum expansion and allow high-speed computer technology to present you the answer. The results of such an approach are often unintelligible. Hence we were drawn to an extended study of the limiting forms of the cross sections, in the low frequency, in the high frequency, and in the high frequency *glory* limits. Finally in chapters 7 and 8 we arrive at the computational level, and agreement with the limiting and qualitative results of chapter 6 give us confidence in our result.

What have we profited? We have developed, and develop still, a variety of techniques of perturbation theory. We have learned much about scattering theory, and we have numerical predictions that one day may allow us to measure the inertial mass of a condensed object, perhaps proving the existence of a black hole.

We close with a brief conjecture on the qualitative features of off-axis black hole scattering. We will discuss each of the principal features seen in the scattering cross sections of axially incident gravitational waves: the forward divergence, the orbital dip, and the backward glory.

The forward divergence in the cross section for axial incidence is part of the overall Newtonian background; it is expected to persist for all energies and all angles of incidence. The reason for this is that the waves scattered through small angles, which consequently contribute to the forward scattering, are those with small $M\omega/l$. The Kerr geometry appears spherically symmetric to such waves, as we saw in chapter 6.

For nonaxial incidence, we would expect that spiralling orbits should

(a)

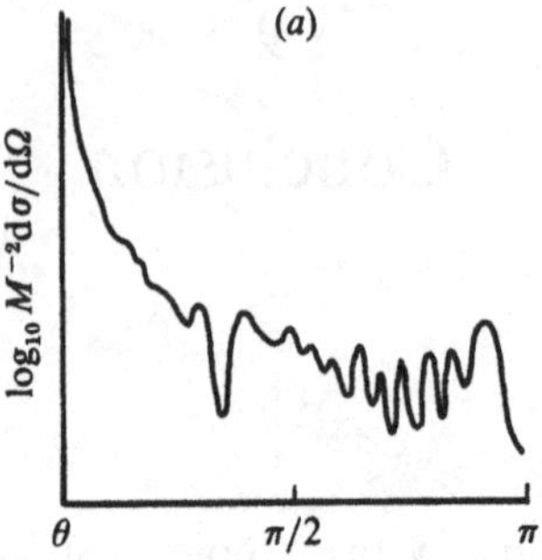

(b)

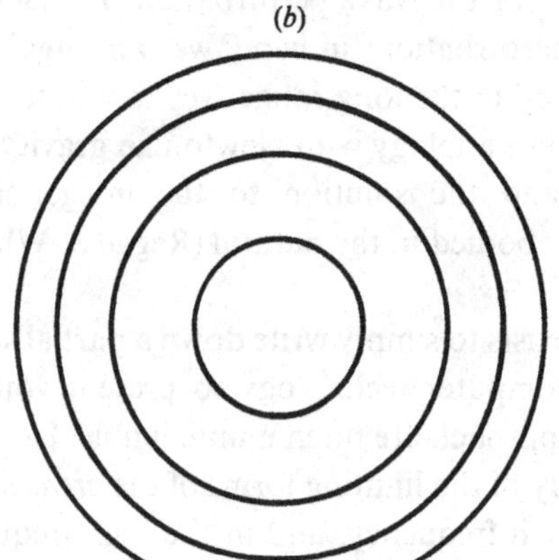

(c)

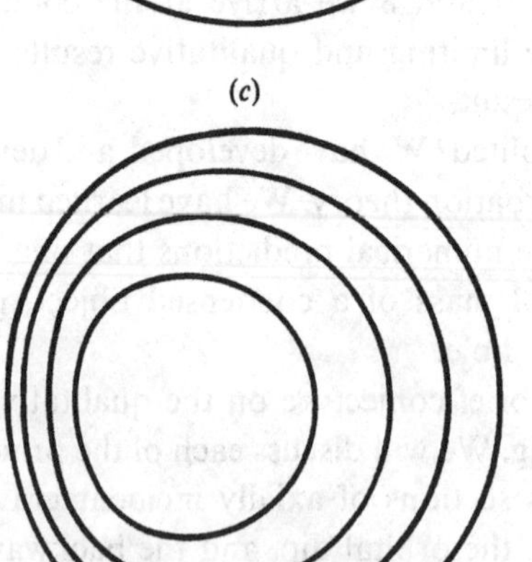

Fig. 9.1.(*a*) A schematic representation of a typical on-axis cross section (compare to fig. 8.17). (*b*) The cross section of fig. 9.1(*a*) used to trace intensity maxima on the celestial sphere. Each of the concentric circles represents a local maximum in the scattering described by the function in fig. 9.1(*a*). (*c*) In the off-axis case we expect distortion of the loci of maximum intensity (Bardeen, 1973).

still be important. In particular the case of equatorial incidence is the only case in Kerr where we would have true single plane spiralling. The axially incident cases calculated in chapter 8 showed evidence of spiralling since the cross sections are a function of the spin of the incident wave.

For the off-axis case, the most important effects must arise because the cross section will no longer be symmetric about the direction of incidence. Instead the cross section for plane waves incident from $\theta = \gamma$, $\phi = 0$ will depend upon both the colatitude θ and the azimuth ϕ. This asymmetry about the incident direction is a consequence of 'frame dragging' by the rotating black hole, and should be strongest for those modes that pass closest to the black hole.

Some idea of this alteration of the cross section may be obtained by considering the high energy (JWKB) limit. Bardeen (1973) calculated the apparent shape of an extreme ($a = M$) Kerr black hole when viewed by an observer in the equatorial plane, if the black hole is between the observer and a source of illumination of greater angular size than the hole. He finds that if the sense of rotation of the hole is taken to be from left to right across the observer's line of sight, the black hole appears displaced to the right and flattened on the left side (see figure 6 of Bardeen (1973)).

The relevance of this distortion is seen as follows. Fig. 9.1(*a*) illustrates schematically a typical cross section from chapter 8 for axial incidence as discussed above. Since the scattering is axially symmetric, we may break the figure at $\theta = \pi/2$ into portions and rotate each about the symmetry axis to yield patterns of bright and dark rings shown in fig. 9.1(*b*). From the preceding discussion we anticipate that these rings will be displaced from the axis of incidence ($\theta = \gamma$, $\phi = 0$) and distorted as in fig. 9.1(*c*). The sketchy arguments given here do not serve to determine whether the bright rings will intersect, but the possibility is not excluded.

APPENDIX A1

Integrals used to express plane waves in terms of spin-weighted spheroidal functions

The asymptotic expansion of plane waves in terms of spin-weighted spheroidal functions is accomplished by evaluating in the limit of large r integrals of the form

$$ {}_sI_l^m(\gamma) = \int \mathrm{d}\Omega \exp\{\mathrm{i}(a_1\chi + (a_3 - m)\phi)\} f(\theta) {}_sS_l^m(\theta; a\tilde{\omega}), \tag{A1} $$

where

$$ \chi = \omega(t - r\sin\gamma\sin\theta\sin\phi - r\cos\gamma\cos\theta). $$

Here γ is the angle between the wave 3-vector and the positive z-axis, the function f is unspecified and ${}_sS_l^m(\theta; a\omega)$ is a spin-weighted angular spheroidal function. The integration extends over the colatitude θ and the azimuth ϕ. In what follows the abbreviation GR *m.mmm.m* refers to a numbered equation from Gradshteyn & Rhyzhik (1965).

The ϕ-integration is easy. If we let

$$ \alpha = -a_1\omega r\sin\gamma\sin\theta, \tag{A2} $$

$$ n = m - a_3 \quad \text{(an integer)}, \tag{A3} $$

a result from standard analysis (Arfken, 1970, p. 490 or GR 8.411.1):

$$ \int_0^{2\pi} \mathrm{d}\phi \exp\{\mathrm{i}[\alpha\sin\phi - n\phi]\} = 2\pi J_n(\alpha) \tag{A4} $$

yields

$$ {}_sI_l^m \simeq \left(\frac{2\pi}{-a_1\omega r}\right)^{1/2} \exp(\mathrm{i}a_1\omega t)\left[\int_0^\pi \frac{\sin\theta f(\theta)}{(\sin\theta\sin\gamma)^{1/2}} \right. $$
$$ \times \exp\{\mathrm{i}[-a_1\omega r\cos(\theta-\gamma) - q]\} {}_sS_l^m(\theta; a\tilde{\omega})\,\mathrm{d}\theta $$
$$ + \int_0^\pi \frac{\sin\theta f(\theta)}{(\sin\theta\sin\gamma)^{1/2}} $$
$$ \left. \times \exp\{\mathrm{i}[-a_1\omega r\cos(\theta+\gamma) + q]\} {}_sS_l^m(\theta; a\tilde{\omega})\,\mathrm{d}\theta \right], $$

where the asymptotic form

$$J_n(\alpha) \underset{|\alpha|\to\infty}{\simeq} \left(\frac{2}{\pi\alpha}\right)^{1/2} \cos\left[\alpha - (n+\tfrac{1}{2})\pi/2\right]$$

$$= \left(\frac{1}{2\pi\alpha}\right)^{1/2} \left[e^{i(\alpha-q)} + e^{i(-\alpha+q)}\right] \tag{A6}$$

and

$$q \equiv (n+\tfrac{1}{2})\pi/2 = (m - a_3 + \tfrac{1}{2})\pi/2 \tag{A7}$$

have been used.

The first integral in (A5) has a point of stationary phase at $\theta = \gamma$, and the second at $\theta = \pi - \gamma$. Since we are evaluating the integrals in the limit of large r, the integrands are assumed to contribute appreciably to the values of the integrals only near the points of stationary phase. We thus consider the behavior of ${}_sS_l^m(\theta; a\omega)$ at $\theta = \gamma$ and $\theta = \pi - \gamma$.

In each of the integrals in (A5) let $h_1(\theta)$, $h_2(\theta)$ denote all functions of θ in the integrand aside from the exponential. Then, under the stationary phase approximation,

$$\begin{aligned} {}_sI_l^m(\gamma) \cong \left(\frac{2\pi}{-a_1\omega r}\right)^{1/2} \exp(ia_1\omega t)\Bigg[e^{-iq} \int_{-\infty}^{\infty} h_1(\gamma) \\ \times \exp\left\{(-ia_1\omega r)\left[1 - \frac{(\theta-\gamma)^2}{2}\right]\right\} d\theta \\ + e^{iq} \int_{-\infty}^{\infty} d\theta\, h_2(\pi-\gamma) \exp\left\{-ia_1\omega r\left[-1 + \frac{(\theta+\gamma-\pi)^2}{2}\right]\right\} d\theta \Bigg] \end{aligned} \tag{A8}$$

where we have evaluated h_1, h_2 at the respective stationary points of their exponents, expanded the exponents about the stationary parts and extended the integration limits to $\pm\infty$.

With appropriate changes of variable both integrals may be put into the form (where n is an integer)

$$I(n) = \int_{-\infty}^{\infty} e^{-\lambda x^2} x^n \, dx \tag{A9}$$

$$\left.\begin{aligned} &= \left(\frac{\pi}{\lambda}\right)^{1/2} \frac{H_n(0)}{(2i\lambda^{1/2})^n} \\ &= \left(\frac{\pi}{\lambda}\right)^{1/2} \left(\frac{-2}{\lambda}\right)^{n/2} \frac{(n-1)!!}{(2i)^n} \quad n \text{ even}, \\ &= 0 \quad n \text{ odd}. \end{aligned}\right\} \tag{A10}$$

Here we have made use of the integral representation of the Hermite polynomials H_n and their values at the origin (GR 8.951, 8.956.6 and 8.956.7).

Thus to leading order in $1/r$

$$ {}_sI_l^m(\gamma) \cong \frac{2\pi}{-a_1\omega r}\Bigg[\exp[\mathrm{i}a_1\omega(t-r)]\exp[\mathrm{i}(a_3-m)\pi/2] \tag{A11}$$

$$\times\exp(-\mathrm{i}\pi/4)f(\gamma)_sS_l^m(\gamma;a\tilde{\omega})\left(\frac{1}{\mathrm{i}}\right)^{1/2}+\exp[\mathrm{i}a_1\omega(t+r)]$$

$$\times\exp[\mathrm{i}(m-a_3)\pi/2]\exp(\pi/4)f(\pi-\gamma)_sS_l^m(\pi-\gamma;a\tilde{\omega})\left(\frac{1}{-\mathrm{i}}\right)^{1/2}\Bigg]$$

where we have restored the definitions of h_1, h_2 and q.

There is some remaining ambiguity in the phase of the terms $(1/\mathrm{i})^{1/2}$, $(1/-\mathrm{i})^{1/2}$. We now show that the correct choice is

$$\left(\frac{1}{\mathrm{i}}\right)^{1/2}=\exp(-\mathrm{i}\pi/4),\quad\left(\frac{1}{-\mathrm{i}}\right)^{1/2}=\exp(+\mathrm{i}\pi/4). \tag{A12}$$

To do so, we should take the limit of (A11) for on-axis scattering. Some care must be exercised, because the large argument limit of the Bessel function (used in (A6)) is incorrect exactly on axis. However, consistency may be obtained by noting essentially that the integral vanishes unless $m=a_3$. We now redo the integral (A1) for the exactly on axis case ($\gamma=0$):

$$ {}_sI_l^m(0)=2\pi\delta_{a_3,m}\int_0^\pi\sin\theta\,\mathrm{d}\theta\,f(\theta)\exp(\mathrm{i}a_1\omega t)\exp(-\mathrm{i}a_1\omega r\cos\theta)_sS_l^m(\theta;a\tilde{\omega}). \tag{A13}$$

We break (A13) into two integrals, one about each point of stationary phase, and write

$$ {}_sI_l^m(\gamma=0)\simeq 2\pi\delta_{a_3m}\exp[\mathrm{i}a_1\omega(t-r)]f(0)_sN_{l;0}^m(a\tilde{\omega}) \tag{A14}$$

$$\times\int_0^\infty\exp(\mathrm{i}a_1\omega r\theta^2/2)\theta^{|m+s|+1}\,\mathrm{d}\theta$$

$$+2\pi\delta_{a_3m}\exp[\mathrm{i}a_1\omega(t+r)]f(\pi)_sN_{l;\pi}^m(a\tilde{\omega})$$

$$\times\int_{-\infty}^\pi\exp[-\mathrm{i}a_1\omega r(\theta-\pi)^2/2](\pi-\theta)^{|m-s|+1}\,\mathrm{d}\theta$$

Changing the variable of integration to $\pi-\theta$ in the second integral in (A14) allows us to express both integrals in the form

$$
\begin{aligned}
I(n) &= \int_0^\infty \mathrm{e}^{-\lambda x^2} x^n \,\mathrm{d}x \qquad\qquad \text{(A15)} \\
&= \left(\frac{\pi}{\lambda}\right)^{n/2} \frac{(n-1)!!}{2(2\lambda)^{n/2}} \qquad n \text{ even}, \\
&= \frac{[(n-1)/2]!}{2\lambda^{(n+1)/2}} \qquad n \text{ odd},
\end{aligned}
$$

where we have used GR 3.461.2 and 3.461.3. Again, retaining the terms only to order $(1/\lambda) \sim (1/r)$ we find that

$$
\begin{aligned}
{}_sI_l^m(\gamma = 0) &= \frac{2\pi \mathrm{i}\delta_{a_3 m}}{a_1 \omega r} \delta_{m-s} \exp\left[\mathrm{i}a_1\omega(t-r)\right] f(0)\,{}_sN_{l:0}^m(a\tilde{\omega}) \\
&\quad - \frac{2\pi \mathrm{i}\delta_{a_3 m}}{a_1 \omega r} \delta_{ms} \exp\left[\mathrm{i}a_1\omega(t+r)\right] f(\pi)\,{}_sN_{l:\pi}^m(a\tilde{\omega}). \qquad \text{(A16)}
\end{aligned}
$$

This agrees with (A11) above at $\gamma = 0$, given the restrictions on a_3, m, imposed by the on-axis situation, and provided the phase choices (A12) are made.

APPENDIX A2

Addition formulae for spin-weighted spherical angular functions

For reference, we derive addition formulae for the spin-weighted spherical angular functions in this section.

We begin with the identification made by Goldberg, *et al.* (1967) of the spin-weighted spherical harmonics

$$ {}_sY_l^m(\theta, \phi) = {}_sS_l^m(\theta)e^{im\phi} \tag{A17}$$

with the matrix elements of the representation of the rotation group of weight l:

$$ {}_sY_l^m(\theta, \phi) = \left(\frac{2l+1}{4\pi}\right)^{1/2} (-1)^m D^l_{-sm}(\phi, \theta, 0). \tag{A18}$$

Equation (A19) is identical to equation (3.11) of Goldberg, *et al.* (1967) where a factor $(-1)^m$ has been inserted in order that the spherical harmonics so defined have the standard Condon–Shortley phase when $s = 0$ (see Arfken, 1970, p. 571). Talman (1968) employs matrix elements related to those of Goldberg, *et al.* by

$$ {}_{\mathrm{T}}D^l_{ms}(-\phi, \theta - \pi, \psi) = (i)^{m-s}(-1)^{l+m}\,{}_{\mathrm{G}}D^l_{-sm}(\phi, \theta, \psi) \tag{A19}$$

where the subscripts 'T' and 'G' denote Talman and Goldberg, respectively. Talman proves sum rules for functions ${}_{\mathrm{T}}d^l_{ms}(\theta)$ defined by

$$ {}_{\mathrm{T}}d^l_{ms}(\pi - \theta) = (-1)^{l+m} \sum_r \frac{[(l-s)!(l+s)!(l-m)!(l+m)!]^{1/2}}{r!(l+m-r)!(l-s-r)!(r+s-m)!}(-1)^{l-s-r} \times \left(\cos\frac{\theta}{2}\right)^{2r+s-m}\left(\sin\frac{\theta}{2}\right)^{2l-2r-s+m} \tag{A20}$$

and related to ${}_{\mathrm{T}}D^l_{ms}$ by

$$ {}_{\mathrm{T}}D^l_{ms}(-\phi, \theta - \pi, 0) = (i)^{m-s}e^{im\phi}\,{}_{\mathrm{T}}d^l_{ms}(\theta - \pi).$$

When we make the identification

$$ (-1)^l\left(\frac{4\pi}{2l+1}\right)^{1/2} {}_sS_l^m(\theta) = {}_{\mathrm{T}}d^l_{ms}(\theta - \pi), \tag{A21}$$

equations (9.20) and (9.70a,b) in Talman (1968, pp. 144, 161) then imply

$$\left(\frac{4\pi}{2l+1}\right)^{1/2}(-1)^l \sum_{j=-l}^{l} {}_jS_l^m(\theta_1)\,{}_sS_l^j(\theta_2) = {}_sS_l^m(\theta_1+\theta_2-\pi), \tag{A22}$$

$$\left(\frac{4\pi}{2l+1}\right)^{1/2}(-1)^l \sum_{j=-l}^{l} (-1)^j\,{}_jS_l^m(\theta_1)\,{}_sS_l^j(\theta_2) = {}_sS_l^m(\theta_1-\theta_2+\pi). \tag{A23}$$

Equations (A20) and (A21) provide an explicit formula for the ${}_sS_l^m(\theta)$ which may be used along with perturbation or continuation techniques to determine the spheroidal functions ${}_sS_l^m(\theta; a\omega)$.

Following Campbell (1971) we take equations (A20) and (A21) as the definition of ${}_sS_l^m(\theta)$ when s, l, and m are half-integral. Thus equations (A22) and (A23) hold for half-integral as well as integral values of s, l, and m.

APPENDIX B

For $h_{\mu\nu}$ given in (3.59), the explicit expressions for h_{ij} in spherical coordinates ($x^1 = r$, $x^2 = \theta$, $x^3 = \phi$) are

$$h_{11} = h[\sin^2\theta(\cos^2\phi - \sin^2\phi\cos^2\gamma) - \cos^2\theta\sin^2\gamma + 2\sin\theta\cos\theta\sin\phi\sin\gamma\cos\gamma]\cos\chi + 2h(\sin^2\theta\cos\phi\sin\phi\cos\gamma - \sin\theta\cos\theta\cos\phi\sin\gamma)\sin\chi, \tag{B1}$$

$$h_{12} = rh[\sin\theta\cos\theta(\cos^2\phi - \sin^2\phi\cos^2\gamma + \sin^2\gamma) + (\cos^2\theta - \sin^2\theta)\sin\phi\cos\gamma\sin\gamma]\cos\chi + rh[2\sin\theta\cos\theta\sin\phi\cos\phi\cos\gamma + (\sin^2\theta - \cos^2\theta)\cos\phi\sin\gamma]\sin\chi, \tag{B2}$$

$$h_{13} = rh[-\sin^2\theta\cos\phi\sin\phi(1 + \cos^2\gamma) + \cos\theta\sin\theta\cos\phi\cos\gamma\sin\gamma]\cos\chi + rh[\sin^2\theta(\cos^2\phi - \sin^2\phi)\cos\gamma + \cos\theta\sin\theta\sin\phi\sin\gamma]\sin\chi, \tag{B3}$$

$$h_{22} = r^2h[\cos^2\theta(\cos^2\phi - \sin^2\phi\cos^2\gamma) - \sin^2\theta\sin^2\gamma - 2\cos\theta\sin\phi\sin\theta\sin\gamma\cos\gamma]\cos\chi + 2r^2h(\cos^2\theta\cos\phi\sin\phi\cos\gamma + \cos\theta\sin\theta\cos\phi\sin\gamma)\sin\chi, \tag{B4}$$

$$h_{23} = -r^2h[\cos\theta\sin\theta\cos\phi\sin\phi(1 + \cos^2\gamma) + \sin^2\theta\cos\phi\cos\gamma\sin\gamma]\cos\chi + r^2h[\sin\theta\cos\theta(\cos^2\phi - \sin^2\phi)\cos\gamma - \sin^2\theta\sin\phi\sin\gamma]\sin\chi \tag{B5}$$

$$h_{33} = r^2h[\sin^2\theta(\sin^2\phi - \cos^2\phi\cos^2\gamma)]\cos\chi - 2r^2h(\sin^2\theta\sin\phi\cos\phi\cos\gamma)\sin\chi, \tag{B6}$$

where

$$\chi = \omega(t - r\sin\gamma\sin\theta\sin\phi - r\cos\gamma\cos\theta) \tag{B7}$$

with γ the angle of inclination to the z-axis.

References

Adler, R. J. & Scheffield, C. (1972). Classification of spacetimes in general relativity, *J. Math. Phys.* **14**, 465.

Arfken, G. (1970). *Mathematical Methods for Physicists*, second edition. Academic Press, New York (875 pages).

Bardeen, J. M. (1973). Timelike and null geodesics in the Kerr metric, in *Black Holes*, Les Houches 1972, ed. C. M. DeWitt & B. S. DeWitt. Gordon & Breach, New York.

Boyer, R. H. & Lindquist, R. W. (1967). Maximal Analytic Extension of the Kerr Metric, *J. Math Phys.* **8**, 265–81.

Breuer, R. A., Ryan, M. P. Jr & Waller, S. (1977). Some properties of spin-weighted harmonics, *Proc. R. Soc. Lond.* A **358**, 71–86.

Brill, D., Chrzanowski, P. L. Pereira, C. M. Fackerell, E. D. & Ipser, J. R. (1972). Solution of the scalar wave equation in Kerr background by separation of variables, *Phys. Rev.* **D5**, 1913.

Brill, D. R. & Wheeler, J. A. (1957). Interaction of neutrinos and gravitational fields, *Rev. Mod. Phys.* **29**, 465–79.

Brillouin, L. (1926). La méchanique ondulatoire de Schrödinger; une méthode générale de résolution par approximations successives, *C. R. Acad. Sci.* **183**, 24–6.

Campbell, W. B. (1971). Tensor and spinor spherical harmonics and the spin-*s* harmonics ${}_sY_l^m(\theta, \phi)$, *J. Math. Phys.* **8**, 1763.

Campbell, W. B. & Morgan, T. (1971). Debye potentials for the gravitational field, *Physica*, **53**, 264.

Chandrasekhar, S. (1975a). On the equations governing the perturbations of the Schwarzschild black hole, *Proc. R. Soc. Lond.* A **343**, 289–98.

Chandrasekhar, S. (1975b). On the equations governing the axisymmetric perturbations of the Kerr black hole, *Proc. R. Soc. Lond.* A **345**, 145–67.

Chandrasekhar, S. (1976). The solution of Dirac's equation in Kerr geometry, *Proc. R. Soc. Lond.* A**349**, 571–5.

Chandrasekhar, S. (1977). On the reflexion and transmission of neutrino waves by a Kerr black hole, *Proc. R. Soc. Lond.* A **352**, 325–38.

Chandrasekhar, S. (1978a). The gravitational perturbations of the Kerr black hole I. The perturbations in the quantities which vanish in the stationary state, *Proc. R. Soc. Lond.* A **358**, 421–39.

Chandrasekhar, S. (1978b). The gravitational perturbations of the Kerr black hole II. The perturbations in the quantities which are finite in the stationary state, *Proc. R. Soc. Lond.* A **358**, 441–65.

Chandrasekhar, S. (1978c). The gravitational perturbations of the Kerr black hole III. Further amplifications, *Proc. R. Soc. Lond.* A**365**, 425–51.

Chandrasekhar, S. (1979a). On the equations governing the gravitational perturbations of the Reissner–Nordstrøm black hole, *Proc. R. Soc. Lond.* A**365**, 453–65.

Chandrasekhar, S. (1979b). An introduction to the theory of the Kerr metric and its perturbations, in *General Relativity: An Einstein Centenary Survey*, ed. S. W. Hawking & W. Israel, pp. 370–453. Cambridge University Press, Cambridge.

Chandrasekhar, S. (1980). The gravitational perturbations of the Kerr black hole IV. The completion of the solution, *Proc. R. Soc. Lond.* A**372**, 475–84.

Chandrasekhar, S. (1983). *The Mathematical Theory of Black Holes*. Clarendon Press, Oxford.
Chandrasekhar, S. & Detweiler, S. L. (1975). The quasi-normal modes of the Schwarzschild black hole, *Proc. R. Soc. Lond.* A **344**, 441–52.
Chandrasekhar, S. & Detweiler, S. L. (1976). On the equations governing the gravitational perturbations of the Kerr black hole, *Proc. R. Soc. Lond.* A **350**, 165–74.
Chandrasekhar, S. & Detweiler, S. L. (1977). On the reflexion and transmission of neutrino waves by a Kerr black hole, *Proc. R. Soc. Lond.* A **352**, 325–38.
Choquet-Bruhat, Y. (1969). Construction de solutions radiatives approchées des equations d'Einstein, *Commun. Math. Phys.* **12**, 16.
Chrzanowski, P. L. (1975). Vector potential and metric perturbations of a rotating black hole, *Phys. Rev.* D **11**, 2042–62.
Chrzanowski, P. L., Matzner, R. A., Sandberg, V. D. & Ryan, M. P. Jr (1976). Zero-mass plane waves in nonzero gravitational backgrounds, *Phys. Rev.* D **14**, 317–26.
Chrzanowski, P. L. & Misner, C. W. (1974). Geodesic synchrotron radiation in the Kerr geometry by the method of asymptotically factorized Green's functions, *Phys. Rev.* D **10**, 1701–21.
Condon, E. U. & Shortley, G. H. (1935). *The Theory of Atomic Spectra*. Cambridge University Press, Cambridge.
Darwin, C. (1959). The gravity field of a particle, I, *Proc. R. Soc. Lond.* A **249**, 180.
Darwin, C. (1961) The gravity field of a particle, II, *Proc. R. Soc. Lond.* A **263**, 39.
Detweiler, S. (1976). On the equations governing the electromagnetic perturbations of the Kerr black hole, *Proc. R. Soc. Lond.* A **349**, 217–30.
Detweiler, S. (1977). On resonant oscillations of a rapidly rotating black hole, *Proc. R. Soc. Lond.* A **352**, 381–95.
Detweiler, S. (1979). Black holes and gravitational waves: perturbation analysis, in *Sources of Gravitational Radiation*, ed. L. Smarr. Cambridge University Press, Cambridge.
Detweiler, S. (1980). The Klein–Gordon equation and rotating black holes, *Phys. Rev.* D **22**, 2323–6.
Detweiler, S. (1982). *Black Holes: Selected Reprints*. AAPT, New York.
DeWitt, B. S. (1964). Dynamical theory of groups and fields, in *Relativity Groups and Topology*, ed. B. DeWitt & C. DeWitt, Gordon & Breach, New York, pp. 587–822.
DeWitt-Morette, C. & Zhang, T. R. (1983). Path integrals and conservation laws, *Phys. Rev.* D **28**, 2503.
Einstein, A. (1915). Erklärung der Perihelbewegung der Merkur aus der allgemeinen Relativitätstheorie, *Sitzungber Preuss. Akad. Wiss., Berlin Sitzber*, **47**, 831–9.
Einstein, A. (1936). Lens-like action of a star by deviation of light in the gravitational field, *Science* **84**, 506.
Flammer, C. (1957). *Spheroidal Wave Functions*. Stanford University Press, Stanford, CA.
Ford, K. W., Hill, D. L., Wakano, M. & Wheeler, J. A. (1959). Quantum effects near a barrier maximum, *Ann. Phys. NY* **7**, 239.
Ford, K. W. & Wheeler, J. A. (1959a). Semiclassical description of scattering, *Ann. Phys. NY* **7**, 259–86.
Ford, K. W. & Wheeler, J. A. (1959b). Application of semiclassical scattering analysis, *Ann. Phys. NY* **7**, 287–322.
Frolov, V. P. (1979). The Newman–Penrose method in the theory of general relativity, in *Problems in the General Theory of Relativity and the Theory of Group Representations*, ed. M. G. Basov, pp. 73–185, Lebedev Physics Inst. Series, Vol. 96, Consultants Bureau, NY.
Futterman, J. A. H. *The scattering of massless plane waves by rotating black holes*, Ph.D. Dissertation, The University of Texas, Austin, Texas 78712.
Futterman, J. A. H. & Matzner, R. A. (1981). Electromagnetic wave scattering by spheroidal objects using a method of spin-weighted harmonics, *Radio Science* **16**, 1303–13.
Futterman, J. A. H. & Matzner, R. A. (1982). Electromagnetic wave scattering by

spheroidal conductors for arbitrary polarization and angle of incidence I. Theory, *Radio Science* **17**, 463–71.

Geroch, R., Held, A. & Penrose, R. (1973). A space-time calculus based on pairs of null directions, *J. Math. Phys.* **17**, 1226–35.

Goldberg, J. N. & Sachs, R. K. (1962). A theorem on Petrov types, *Acta Physica Polonica* **22**, 13–23.

Goldberg, J. N., MacFarlane, A. J. Newman, E. T. Rohrlich, F. & Sudarshan, E. C. G. (1967). Spin-*s* spherical harmonics and ð, *J. Math. Phys.* **8**, 2155–61.

Goldstein, H. (1950). *Classical Mechanics*. Addison Wesley, Reading, MA.

Gordon, W. (1928). Über den Stoss zweier Punktladungen nach der Wellenmechanic, *Z. Phys.* **48**, 180.

Gottfried, K. (1966). *Quantum Mechanics*. Benjamin, Menlo Park.

Gradshteyn, I. S. & Ryzhik, I. M. (1965). *Table of Integrals, Series and Products*. Academic Press, New York.

Güven, R. (1977). Wave mechanics of electrons in Kerr geometry, *Phys. Rev.* D **16**, 1706–11.

Güven, R. (1980). Black holes have no superhair, *Phys. Rev.* D **22**, 2327–30.

Hammermesh, M. (1962). *Group Theory*. Addison Wesley, Reading, MA.

Handler, F. A. (1979). *Vacuum black hole gravitational wave cross sections*, Ph.D. Dissertation. The University of Texas at Austin, Austin, TX 78712.

Handler, F. A. & Matzner, R. A. (1980). Gravitational wave scattering, *Phys. Rev.* D **22**, 2331–48.

Hawking, S. W. & Ellis, G. F. R. (1973). *The Large Scale Structure of Space-Time*. Cambridge University Press, Cambridge (391 pages).

Hildreth, W. W. (1964). Ph.D. Thesis. Princeton University.

Hjellming, R. M. (1973). Radio variability of HDE 226868 (Cygnus X-1), *Astrophys. J. Lett.* **182**, 229.

Isaacson, R. A. (1968). Gravitational radiation in the limit of high frequency, II. Nonlinear terms and the effective stress tensor, *Phys. Rev.* **166**, 1272–80.

Jackson, A. A. (1985). Note on intensity gain by a gravitational lens, *Am. J. Phys.* **52**, 372–3.

Jackson, J. D. (1962). *Classical Electrodynamics*. Wiley, New York.

Jeffries, H. (1923). On certain approximate solutions to linear differential equations of the second order, *Proc. Lond Math. Soc.* **2**, 428–36.

Jones, B. F. (1976). Gravitational deflection of light: solar eclipse of 30 June 1973, II. Plate reductions, *Astron. J.* **81**, 455–63.

Kerr, R. P. (1963). Gravitational field of a spinning mass as an example of algebraically special metrics, *Phys. Rev. Lett.* **11**, 237–8.

Kinnersley, W. (1969). Type D vacuum metrics, *J. Math. Phys.* **10**, 1195–203.

Kramers, H. A. (1926). Wellenmechanik and Halbzahlige Quantisierung, *Z. Phys.* **39**, 828–40.

Lightman, A. P. & Shapiro, S. L. (1976). Polarization of X-rays from Cygnus X-1: a test of the accretion disk model, *Astrophys. J.* **203**, 701.

Martelli, M. & Treves, A. (1977). Absence of superradiance of a Dirac field in a Kerr background, *Phys. Rev.* D **10**, 3060–61.

Mashoon, B. (1973). Scattering of electromagnetic radiation from a black hole, *Phys. Rev.* D **7**, 2807–13.

Mashoon, B. (1974). Electromagnetic scattering from a black hole and the glory effect, *Phys. Rev.* D **10**, 1059.

Mathews, J. & Walker, R. L. (1970). *Mathematical Methods of Physics*. Benjamin, Menlo Park, CA.

Matzner, R. A. (1968). Scattering of massless scalar waves by a Schwarzschild ‘singularity’, *J. Math. Phys.* **9**, 163–70.

Matzner, R. A. (1976). Low frequency-limit conversion cross-sections for charged black holes, *Phys. Rev.* D **14**, 3274–80.

Matzner, R. A. & Ryan, M. P. Jr (1977). Low-frequency limit of gravitational scattering, *Phys. Rev.* D **16**, 1636–42.

Matzner, R. A. & Ryan, M. P. Jr (1978). Scattering of gravitational radiation from vacuum black holes, *Astrophys. J. Suppl.* **36**, 451–81.

Matzner, R. A., DeWitt-Morette, C., Nelson, B. & Zhang, T.-R. (1985). Glory Scattering by black holes, *Phys. Rev.* D **31**, 1869.

Merzbacher, E. (1971). *Quantum Mechanics*, second edition, Wiley, New York.

Messiah, A. (1958). *Quantum Mechanics*, Wiley, New York, (1136 pages).

Misner, C. W., Thorne, K. S. & Wheeler, J. A. (1973). *Gravitation*, Wiley, New York.

Moncrief, V. (1975). Gauge-invariant perturbations of Reissner–Nordstrom black holes, *Phys. Rev.* D **12**, 1526.

Nelson, B. & DeWitt-Morette, C. (1984). Glories – and other degenerate critical points of the Action, *Phys. Rev.* D **29**, 1663–8.

Newman, E., Couch, E., Chinnapared, K., Exton, A., Prakash, A. & Torrence, R. (1965). Metric of a rotating charged mass, *J. Math. Phys.* **6**, 918–19.

Newman, E. & Penrose, R. (1962). An approach to gravitational radiation by a method of spin coefficients. *J. Math. Phys.* **3**, 566–78.

Newton, R. G. (1966). *Scattering Theory of Waves and Particles*. McGraw-Hill, New York.

Nordstrøm, G. (1918). On the energy of the gravitational field in Einstein's theory, *Proc. Kon. Ned. Akad. Wet.* **20**, 1238–45.

Penrose, R. (1960). A spinor approach to general relativity, *Ann. Phys.* **10**, 171–201.

Penrose, R. (1968). Structure of spacetime, in *Battelle Rencontres*, ed. C. M. DeWitt & J. A. Wheeler. W. A. Benjamin, New York, pp. 121–235.

Penrose, R. (1969). Gravitational collapse: the role of general relativity, *Nuovo Cimento Ser. I*, **1**, 252–76.

Peters, P. C. (1976). Differential cross sections for weak-field gravitational scattering, *Phys. Rev.* D **13**, 775.

Petrov, A. Z. (1955). *Dokl. Akad. Nauk. SSSR* **105**, 905.

Pirani, F. A. E. (1965). Introduction to gravitational radiation theory, in *Lectures on General Relativity*, ed. A. Trautman, F. A. E. Pirani & H. Bondi. Prentice Hall, Englewood Cliffs, NJ.

Press, W. H. & Teukolsky, S. A. (1973). Perturbations of a rotating black hole II. Dynamical stability of the Kerr metric, *Astrophys. J.* **185**, 649–73.

Reasenberg, R. D., Shapiro, I. I., MacNeil, P. E., Goldstein, R. B., Breidenthal, J. C., Brenkle, J. P., Cain, D. L., Kaufman, T. M., Komarek, T. A. & Zygielbaum A. F. (1979). *Astrophys. J.* **234**, L219–L221.

Regge, T. & Wheeler, J. A. (1957). Stability of a Schwarzschild singularity, *Phys. Rev.* **108**, 1063–9.

Reissner, H. (1916). Über die Eigengravitation des Elektrischen Feldes nach der Einsteinschen Theorie, *Ann. Phys.* **50**, 106–20.

Rose, M. E. (1961). *Relativistic Electron Theory*. John Wiley, New York.

Rosenman, J. G. (1971). *Dirac electrons in a gravitational field*, Ph.D Thesis, The University of Texas at Austin, Austin, Texas 78712.

Sachs, R. K. (1964). Gravitational radiation, in *Relativity; Groups and Topology*, ed. C. M. DeWitt & B. S. DeWitt. Gordon & Breach, New York.

Sakurai, J. J. (1967). *Advanced Quantum Mechanics*, Addison-Wesley, Reading, MA.

Sanchez, N. (1976). Scattering of scalar waves from a Schwarzschild black hole, *J. Math. Phys.* **17**, 688–92.

Sanchez, N. (1977). Wave scattering theory and the absorption problem for a black hole, *Phys. Rev.* D **16**, 937–45.

Sanchez, N. (1978a). Absorption and emission spectra for Schwarzschild black hole, *Phys. Rev.* D **18**, 1030–6.
Sanchez, N. (1978b). Elastic scattering of waves by a black hole, *Phys. Rev.* D **18**, 1798–804.
Schiff, L. (1968). *Quantum Mechanics*, 3rd edition, McGraw-Hill, New York.
Schwarzschild, K. (1916). Über das Gravitationsfeld eines Massenpunketes nach der Einsteinschen Theorie, *Sitzber. Deut. Acad. Wiss. Berlin, Kl. Math.-Phys. Tech.*, 189–96.
Schweber, S. S. (1961). *An Introduction to Relativistic Quantum Field Theory.* Harper & Row, New York.
Soldner, J. (1801). *Astron. Jahrbuch f. d. Jahr 1804*, p. 161 (Berlin). This article was reprinted with comments of an anti-Einstein character by P. Lenard (1921). *Ann. Phys. Lpz.* **65**, 593.
Starobinsky, A. A. & Churilov, S. M. (1973). Amplification of electromagnetic and gravitational waves scattered by a rotating black hole, *Zh. Eksp. Theor. Fiz.* **65**, 3 (also (1974) *Sov. Phys. JETP* **38**, 1–5).
Stewart, J. & Walker, M. (1973). Black holes: the outside story, in *Astrophysics* (Springer Tract on Modern Physics, Vol. 69), Springer, Berlin.
Talman, J. D. (1968). *Special functions.* W. A. Benjamin, New York.
Tananbaum, H., Gursky, H., Kellogg, E., Giacconi, R. & Jones, C. (1972). Observations of a correllated X-ray–Radio transition in Cygnus X-1, *Astrophys. J. Lett.* **177**, 25.
Teukolsky, S. A. (1972). Rotating black holes; separable wave equations for gravitational and electromagnetic perturbations, *Phys. Rev. Lett.* **29**, 1114–18.
Teukolsky, S. A. (1973). Perturbations of a rotating black hole I. Fundamental equations for gravitational, electromagnetic and neutrino-field perturbations, *Astrophys. J.* **185**, 635–47.
Teukolsky, S. A. & Press, W. H. (1974). Perturbations of a rotating black hole III. Interaction of the hole with gravitational and electromagnetic radiation, *Astrophys. J.* **193**, 443–61.
Unruh, W. G. (1973). Separability of the neutrino equations in a Kerr background, *Phys. Rev. Lett.* **31**, 1265–7.
Unruh, W. G. (1974). Second quantization in the Kerr metric, *Phys. Rev.* D **10**, 3194–205.
Unruh, W. G. (1976). Absorption cross section of small black holes, *Phys. Rev.* D **14**, 3251–9.
Vishveshwara, C. V. (1970). Stability of the Schwarzschild metric, *Phys. Rev.* D **1**, 2870–9.
Wald, R. M. (1973). On perturbations of a Kerr black hole, *J. Math. Phys.* **14**, 1453.
Wald, R. M. (1978). Construction of solutions of gravitational, electromagnetic, or other perturbation equations from solutions of decoupled equations, *Phys. Rev. Lett.* **41**, 203–6.
Wasserstrom, E. (1972). A new method for solving eigenvalue problems, *J. Comput. Phys.* **9**, 53–74.
Weber, J. (1970). Gravitational radiation experiments, *Phys. Rev. Lett.* **24**, 276.
Weinberg, S. (1972). *Gravitation and Cosmology.* Wiley, New York.
Wentzel, G. (1926). Eine Verallgemeinerung der Quantenbedingungen für die Zwecke der Wellenmechanik, *Z. Phys.* **38**, 518–29
Westervelt, P. J. (1971). Scattering of electromagnetic and gravitational waves by a static gravitational field: comparison between the classical (general relativistic) and quantum field theoretic results, *Phys. Rev.* D **3**, 2319.
Young, P. J., Gunn, J. E., Kristen, J. A., Oke, J. B. & Westpool, J. A. (1980). *Astrophys. J.* **241**, 507.
Zel'dovich, Ya. B. (1971). Generation of waves by a rotating body, *JETP* **14**, 180–1.
Zerilli, F. J. (1970). Effective potential for even-parity Regge–Wheeler gravitational perturbation equations, *Phys. Rev. Lett.* **24**, 737–8.

Index